AF389014

Advancements in Catalytic Conversion of Biomass into Biofuels and Chemicals

Advancements in Catalytic Conversion of Biomass into Biofuels and Chemicals

Editors

Tae Hyun Kim
Chang Geun Yoo

MDPI • Basel • Beijing • Wuhan • Barcelona • Belgrade • Manchester • Tokyo • Cluj • Tianjin

Editors
Tae Hyun Kim
Hanyang University
Korea

Chang Geun Yoo
State University of New York College of Environmental Science and Forestry
USA

Editorial Office
MDPI
St. Alban-Anlage 66
4052 Basel, Switzerland

This is a reprint of articles from the Special Issue published online in the open access journal *Energies* (ISSN 1996-1073) (available at: https://www.mdpi.com/journal/energies/special_issues/ biochemical_process).

For citation purposes, cite each article independently as indicated on the article page online and as indicated below:

LastName, A.A.; LastName, B.B.; LastName, C.C. Article Title. *Journal Name* **Year**, *Volume Number*, Page Range.

ISBN 978-3-03943-715-3 (Hbk)
ISBN 978-3-03943-716-0 (PDF)

Cover image courtesy of Chang Geun Yoo.

Contents

About the Editors

Tae Hyun Kim is currently a professor of the Department of Materials Science & Chemical Engineering at Hanyang University, and he was formerly a faculty member at Iowa State University and Kongju National University. He received his B.S. degree in Chemical Engineering from Hanyang University in 1994. He received his Ph.D. degree in Chemical Engineering in 2004. He has worked at two companies, namely Samsung Engineering and LG Chemical, and at USDA-ARS-ERRC.

Chang Geun Yoo is currently an assistant professor in the Department of Chemical Engineering at State University of New York College of Environmental Science and Forestry. He received his B.S. and M.S. degrees in Chemical Engineering from Hanyang University and received his Ph.D. degree in Agricultural and Biosystems Engineering from Iowa State University in 2012. After graduation, he worked as a postdoc research associate at the University of Wisconsin–Madison and Oak Ridge National Laboratory. He started his current position at SUNY ESF in 2018.

Editorial

Advancements in Catalytic Conversion of Biomass into Biofuels and Chemicals

Chang Geun Yoo [1],* and Tae Hyun Kim [2],*

[1] Department of Chemical Engineering, State University of New York College of Environmental Science and Forestry, Syracuse, NY 13210, USA
[2] Department Materials Science and Chemical Engineering, Hanyang University, Ansan, Gyeonggi-do 15588, Korea
* Correspondence: cyoo05@esf.edu (C.G.Y.); hitaehyun@hanyang.ac.kr (T.H.K.)

Received: 13 October 2020; Accepted: 14 October 2020; Published: 19 October 2020

Abstract: The shortage of resources and increasing climate changes have brought the need for sustainable and renewable resources to people's attention. Biomass is an earth-abundant material and has great potential as a feedstock for alternative fuels and chemicals. For the effective utilization of biomass, this biopolymer has to be depolymerized and transformed into key building blocks and/or the targeted products, and biological or chemical catalysts are commonly used for the rapid and energy-efficient reactions. This Special Issue introduces recent advances in the catalytic conversion of biomass into biofuels and value-added products.

Keywords: biomass; catalysts; solvents

Over the past several decades, petroleum has been used as an essential resource in our lives. However, the rapid increase in the consumption of petroleum-based resources and environmental issues such as greenhouse gas emissions and climate change demand the use of alternative resources. Compared to other renewable sources, such as solar, wind, hydro, and geothermal, biomass has a broader spectrum of applications (e.g., biofuels, biochemicals, biomaterials). However, in order to utilize biomass, its rigid cell wall structures need to be deconstructed and selectively fractionated in advance. The cell walls of biomass are mainly composed of cellulose, hemicellulose, and lignin with other minor components such as extractives, ash, and protein. Recalcitrance factors, including crystalline cellulose, acetylated xylan, lignin, lignin-carbohydrate complex, and inorganic components, must be removed or at least reduced to some extent for effective biomass conversion [1,2].

Pretreatment is designed to eliminate these factors using various solvents and catalysts. Hydrothermal, acid, alkaline, organosolv and other methods have been applied to improve the biological conversion of cellulose and hemicellulose into biofuels and biochemicals [3–6]. Notable enhancements in enzymatic hydrolysis and fermentation were reported with these pretreatments. In addition, robust and effective biocatalysts (i.e., enzymes) and microorganisms have solved many technical problems. Unfortunately, due to the low price of petroleum-based resources these days, these bio-products are not cost-competitive by themselves.

Further enhancement of the efficiency of biomass conversion process and valorization of by-products such as lignin are the possible solutions to overcome this challenge. Several solvents, such as ionic liquids, deep eutectic solvents, molten salt hydrates, and cellulose-derived solvents, have been designed to improve the biomass conversion [7–10]. These solvents provided similar or better biomass conversion under milder conditions compared to the traditional pretreatments. Many catalysts were also designed for improving other key reactions such as hydrolysis, isomerization, hydrogenation, and oxidation of carbohydrates and lignin [11–14]. In particular, solid heterogeneous catalysts have much better catalyst recyclability and product separation than homogeneous catalysts and biocatalysts [15,16].

Fractionation strategies for co-utilizing hemicellulose and lignin have also been studied. The fractionation process aims to separate the unnecessary components from the main product stream with high purity and yield. Therefore, finding and applying a selective as well as an effective catalyst and solvent are crucial. A biphasic solvent system [10] or co-solvent system mixed with an organic and aqueous solution [17] specifically increased the recovery of hydrophobic components (e.g., lignin) and degradation products (e.g., furfural and HMF) by preventing unwanted side reactions in aqueous environments. In addition to the high separation efficiency of each component, fractionation can preserve the quality of the separated components by minimizing their condensation and formation of pseudo-lignin.

Both solvents and catalysts play an important role in biomass conversion. This Special Issue introduces a recent catalytic biomass conversion and upgrade approaches. The published articles provide a clue to solve the technical and economic challenges in biomass utilization.

Funding: This article received no external funding.

Conflicts of Interest: The authors declare no conflict of interest.

References

1. Yoo, C.G.; Yang, Y.; Pu, Y.; Meng, X.; Muchero, W.; Yee, K.L.; Thompson, O.A.; Rodriguez, M.; Bali, G.; Engle, N.L.; et al. Insights of biomass recalcitrance in natural *Populus trichocarpa* variants for biomass conversion. *Green. Chem.* **2017**, *19*, 5467–5478. [CrossRef]
2. Yoo, C.G.; Meng, X.; Pu, Y.; Ragauskas, A.J. The critical role of lignin in lignocellulosic biomass conversion and recent pretreatment strategies: A comprehensive review. *Bioresour. Technol.* **2020**, *301*, 122–784. [CrossRef] [PubMed]
3. Yoo, C.G.; Nghiem, N.P.; Hicks, K.B.; Kim, T.H. Maximum production of fermentable sugars from barley straw using optimized soaking in aqueous ammonia (SAA) pretreatment. *Appl. Biochem.* **2013**, *169*, 2430–2441. [CrossRef] [PubMed]
4. Cao, S.; Pu, Y.; Studer, M.; Wyman, C.; Ragauskas, A.J. Chemical transformations of *Populus trichocarpa* during dilute acid pretreatment. *RSC Adv.* **2012**, *2*, 10925–10936. [CrossRef]
5. Pan, X.; Xie, D.; Yu, R.W.; Saddler, J.N. The bioconversion of mountain pine beetle-killed lodgepole pine to fuel ethanol using the organosolv process. *Biotechnol. Bioeng.* **2008**, *101*, 39–48. [CrossRef] [PubMed]
6. Ko, J.K.; Kim, Y.; Ximenes, E.; Ladisch, M.R. Effect of liquid hot water pretreatment severity on properties of hardwood lignin and enzymatic hydrolysis of cellulose. *Biotechnol. Bioeng.* **2015**, *112*, 252–262. [CrossRef] [PubMed]
7. Wang, Y.; Meng, X.; Jeong, K.; Li, S.; Leem, G.; Kim, K.H.; Pu, Y.; Ragauskas, A.J.; Yoo, C.G. Investigation of a lignin-based deep eutectic solvent using p-hydroxybenzoic acid for efficient woody biomass conversion. *ACS Sustain. Chem. Eng.* **2020**, *8*, 12542–12553. [CrossRef]
8. Liu, E.; Li, M.; Das, L.; Pu, Y.; Frazier, T.; Zhao, B.; Crocker, M.; Ragauskas, A.J.; Shi, J. Understanding lignin fractionation and characterization from engineered switchgrass treated by an aqueous ionic liquid. *ACS Sustain. Chem. Eng.* **2018**, *6*, 6612–6623. [CrossRef]
9. Meng, X.; Pu, Y.; Li, M.; Ragauskas, A.J. A biomass pretreatment using cellulose-derived solvent Cyrene. *Green. Chem.* **2020**, *22*, 2862–2872. [CrossRef]
10. Yoo, C.G.; Zhang, S.; Pan, X. Effective conversion of biomass into bromomethylfurfural, furfural, and depolymerized lignin in lithium bromide molten salt hydrate of a biphasic system. *RSC Adv.* **2017**, *7*, 300–308. [CrossRef]
11. Yoo, C.G.; Li, N.; Swannell, M.; Pan, X. Isomerization of glucose to fructose catalyzed by lithium bromide in water. *Green. Chem.* **2017**, *19*, 4402–4411. [CrossRef]
12. Shuai, L.; Pan, X. Hydrolysis of cellulose by cellulase-mimetic solid catalyst. *Energy Environ. Sci.* **2012**, *5*, 6889–6894. [CrossRef]

13. Nagy, M.; David, K.; Britovsek, G.J.; Ragauskas, A.J. Catalytic hydrogenolysis of ethanol organosolv lignin. *Holzforschung* **2009**, *63*, 513–520. [CrossRef]

14. Schutyser, W.; Kruger, J.S.; Robinson, A.M.; Katahira, R.; Brandner, D.G.; Cleveland, N.S.; Mittal, A.; Peterson, D.J.; Meilan, R.; Román-Leshkov, Y. Revisiting alkaline aerobic lignin oxidation. *Green. Chem.* **2018**, *20*, 3828–3844. [CrossRef]

15. Behling, R.; Valange, S.; Chatel, G. Heterogeneous catalytic oxidation for lignin valorization into valuable chemicals: What results? What limitations? What trends? *Green Chem.* **2016**, *18*, 1839–1854. [CrossRef]

16. Jiang, L.; Guo, H.; Li, C.; Zhou, P.; Zhang, Z. Selective cleavage of lignin and lignin model compounds without external hydrogen, catalyzed by heterogeneous nickel catalysts. *Chem. Sci.* **2019**, *10*, 4458–4468. [CrossRef] [PubMed]

17. Cai, C.M.; Zhang, T.; Kumar, R.; Wyman, C.E. THF co-solvent enhances hydrocarbon fuel precursor yields from lignocellulosic biomass. *Green. Chem.* **2013**, *15*, 3140–3145. [CrossRef]

Publisher's Note: MDPI stays neutral with regard to jurisdictional claims in published maps and institutional affiliations.

Article

Improvement of Organosolv Fractionation Performance for Rice Husk through a Low Acid-Catalyzation

Tae Hoon Kim [1,2], Hyun Jin Ryu [1] and Kyeong Keun Oh [1,3,*

[1]   R&D Center, SugarEn Co., Ltd., Yongin, Gyeonggi-do 16890, Korea; thkim@sugaren.co.kr (T.H.K.); hjryu@sugaren.co.kr (H.J.R.)
[2]   Department of Materials Science and Chemical Engineering, Hanyang University, Ansan, Gyeonggi-do 15588, Korea
[3]   Department of Chemical Engineering, Dankook University, Youngin, Gyeonggi-do 16890, Korea
*   Correspondence: kkoh@dankook.ac.kr; Tel.: +82-31-8005-3548

Received: 18 April 2019; Accepted: 9 May 2019; Published: 11 May 2019

Abstract: For the effective utilization of rice husk, organosolv fractionation was investigated to separate three main components (glucan, xylose, and lignin) with low acid concentration. Reaction temperatures of 170–190 °C, ethanol concentrations of 50%–70% (v/v), and sulfuric acid concentrations of 0%–0.7% (w/v) were investigated, with the reaction time and liquid-to-solid ratio kept constant at 60 min and 10, respectively. The fractionation conditions for the efficient separation into the three components of rice husk were determined to be 180 °C, 60% (v/v) of ethanol, and 0.25% (w/v) of sulfuric acid. Under these fractionation conditions, 86.8% of the xylan and 77.5% of the lignin were removed from the rice husk, and xylose and lignin were obtained from the liquid in 67.6% and 49.8% yields, respectively. The glucan digestibility of the fractionated rice husk was 85.2% with an enzyme loading of 15 FPU (filter paper unit) of cellulase per g-glucan.

Keywords: biomass; xylan; lignin; cellulose; pretreatment

---

1. Introduction

World rice production is 685 million tons per year, and 137 million tons of rice husk (RH) is generated [1]. Despite the enormous amount of RH, most of it is burned or buried in the ground because of lax environmental standards and technological limitations [2]. However, researchers have been attempting to develop efficient uses for RH. Because of its high ash content, RH has been studied in fields such as absorbents, coatings, pigments, the cement industry, insulators, rubber, and electronics [3]. However, these studies have focused only on the ash content of RH, which cannot be considered an efficient use. The composition of RH differs by location but typically includes 49.5% to 64.2% carbohydrates, including cellulose and hemicellulose, and 13.5% to 40.2% lignin [4]. In general, carbohydrates and lignin are sources of high-value-added materials in the biorefinery field. For example, pure cellulose can be converted into fibers or energy and hemicellulose can be converted into high-value-added materials such as furfural, succinic acid, and xylooligosaccharide [5]. Lignin can be used as phenolic platform chemicals such as catechols, cresols, or hydrocarbon precursors [6]. Therefore, the fractionation of lignocellulosic biomass into major components such as cellulose, hemicellulose, and lignin is an effective method to use RH [5].

Different catalysts, including acids, alkalis, organic solvents, and ionic liquids, have been used to separate the major components of lignocellulosic biomass [7]. Among these chemicals, organic solvents offer numerous advantages. Organic solvent (organosolv) fractionation can separate RH into three major components in a single process and therefore assist downstream processing, e.g., enzymatic hydrolysis [8]. When ethanol (EtOH) is used as an organic solvent, it can dissolve hemicellulose and lignin in the liquid hydrolyzate while leaving a high content of cellulose in the residual solid. The removal of hemicellulose and lignin from biomass improves the enzymatic digestibility of cellulose by increasing enzyme accessibility [9]. The lignin extracted into the liquid can be easily precipitated by exploiting the difference in EtOH solubility, and high-quality, high-purity lignin is precipitated [10]. Additional advantages of organosolv-precipitated lignin include a low molecular weight, uniform molecular weight distribution, hydrophobicity, and a low glass transition temperature, which makes the precipitated lignin easy to use in various applications, as previously mentioned [11]. Furthermore, EtOH can be easily recovered and recycled, which is economically advantageous [12]. Although EtOH is known to exhibit high solubilization of lignin, it is generally used in conjunction with an added acid catalyst. The use of an acid catalyst not only leads to a mild reaction but also decomposes the carbohydrate-lignin complex more easily than when an acid catalyst is not used [9,13]. Therefore, the use of an appropriate acid catalyst can improve the efficiency of the fractionation process.

The purpose of this study is to effectively extract xylose and lignin from RH and to improve the enzymatic digestibility of fractionated solids. The EtOH organosolv process was evaluated at various independent variables (reaction temperatures, EtOH concentrations, and sulfuric acid (SA) concentrations). The effects of independent variables were determined in order to selectively extract the desired component in one stage process. In addition, the chemical characteristics of acid-free and acid-catalyzed organosolv precipitated lignin were compared.

2. Results and Discussion

2.1. Organosolv Fractionation of RH

2.1.1. Organosolv Fractionation with Reaction Temperature and EtOH Concentration

Figure 1 presents the compositions of fractionated solid after organosolv fractionation (acid-free) under various reaction conditions. The effects of different reaction temperatures (170–190 °C) and EtOH concentrations (50%–70%) were investigated. As shown in Figure 1, the glucan and xylan contents in the fractionated solid were better retained at high EtOH concentrations (70%) than at low EtOH concentration (50%) at all of the investigated reaction temperatures (170–190 °C). This phenomenon is attributed to the hydrolysis reaction of water at different EtOH–water ratios. That is, a low EtOH concentration means a high concentration of hydronium ions, which would increase the reaction severity [14,15]. The pH at initial EtOH concentrations of 50%, 60%, and 70% were 6.24, 6.25, and 6.25, respectively; however, after reaction (170 °C, 60 min, liquid-to-solid (L/S) ratio of 10), the pH levels of 50%, 60%, and 70% EtOH solutions were 3.32, 4.12, and 4.89, respectively. These results are consistent with the results of the liquid hydrolyzate analysis shown in Figure 2. As the EtOH concentration (50%–70%) and reaction temperature (170–190 °C) were increased, smaller amounts of sugars (glucose and xylose) and byproducts (formic acid, acetic acid, levulinic acid, 5-HMF, and furfural) were released in the liquid hydrolyzate. As byproducts (except for acetic acid) are produced by the decomposition of cellulose and hemicellulose, they are generally used as an indicator of reaction severity [16]. In particular, the 5-HMF was generated under the most severe reaction conditions (190 °C, 50%) via hexose decomposition. Under these conditions (190 °C, 50%), the lignin extraction yield was only 46.0%; thus, large amounts of lignin remained in the solids.

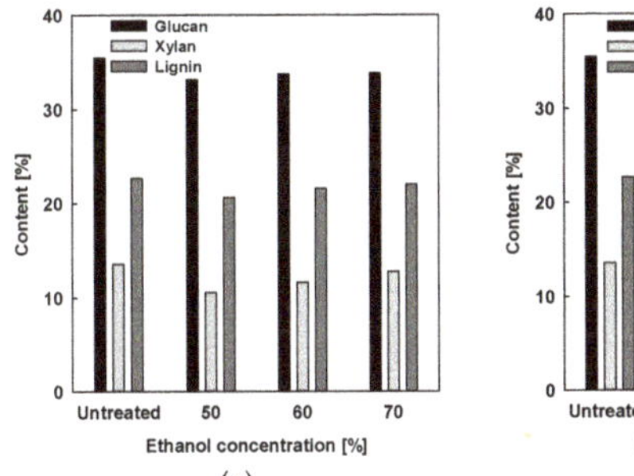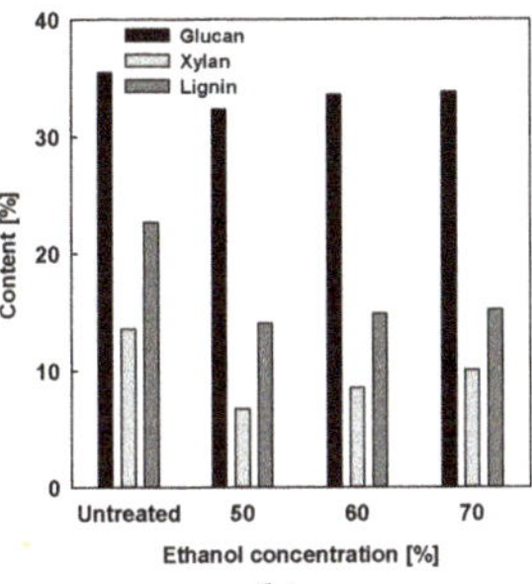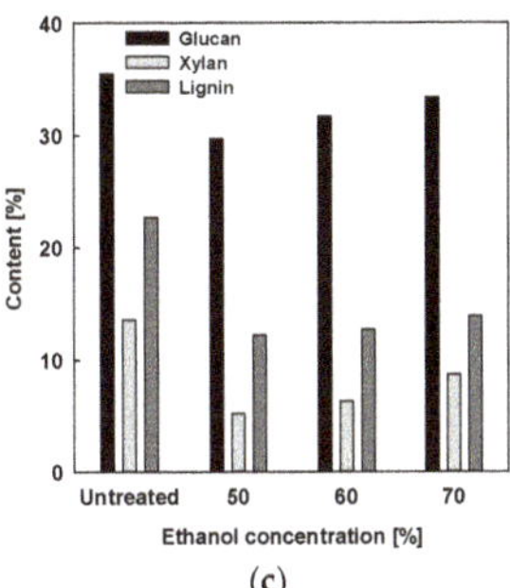

Figure 1. Effects of reaction temperature and EtOH concentration on the chemical compositions of the raw and fractionated solids of rice husk: (**a**) 170 °C, (**b**) 180 °C, and (**c**) 190 °C. Note: reaction conditions: 170–190 °C, 50%–70% EtOH concentration.

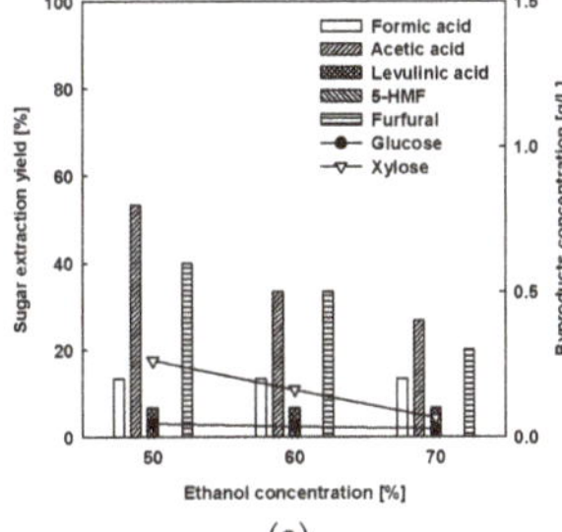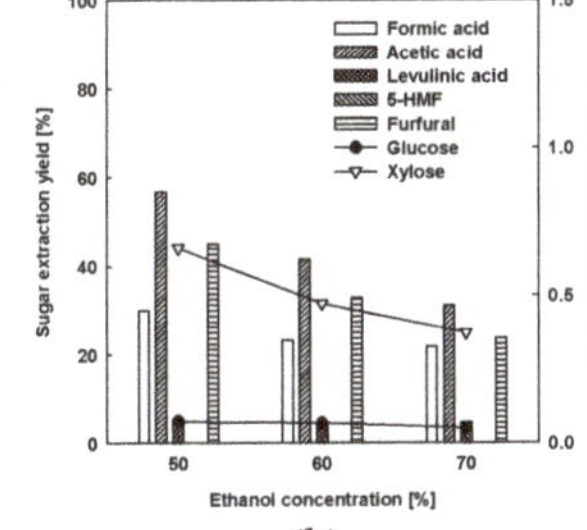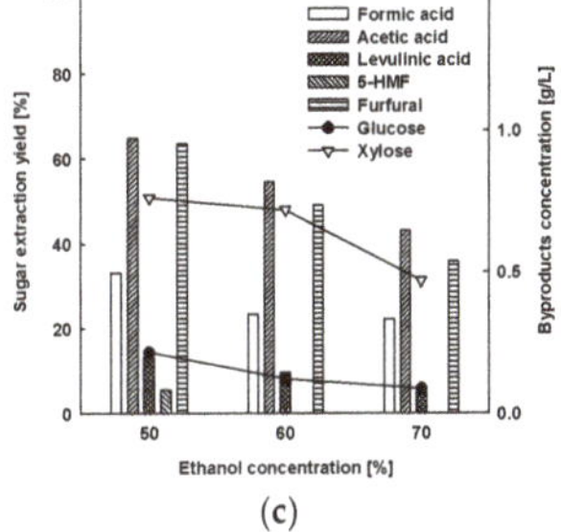

Figure 2. Effects of reaction temperature and EtOH concentration on the sugar extraction yield and byproducts concentration of liquid hydrolyzate: (**a**) 170 °C, (**b**) 180 °C, and (**c**) 190 °C. Note: reaction conditions: 170–190 °C, 50%–70% EtOH concentration.

The SA was added as an acid catalyst to improve the organosolv fractionation performance. Figure 3 presents the effects of reaction temperature (180 °C, 190 °C) and EtOH concentration (50%–70%) on the chemical composition of the fractionated solid. As shown in Figure 3, the xylan content decreased under all of the investigated reaction conditions. In particular, xylan was not detected under the conditions of 190 °C and 50% or 60% EtOH concentration. However, as shown in Figure 4, xylose extracted from the RH was not present in the liquid hydrolyzate. The formation of furfural indicates decomposition of the pentose sugar (xylose), which in turn indicates an increase in the severity of the reaction conditions. As shown in Figure 4, the xylose extraction yield of liquid hydrolyzate was greater at 180 °C than at 190°C, which means that xylose was decomposed into furfural because of the severe conditions (190 °C). However, as shown in Figure 3, the glucan content (88.4%–90.3%) was well preserved in the fractionated RH at 180 °C but was somewhat lower (76.1%–77.4%) at 190 °C. As previously mentioned, glucose was considered to be decomposed into byproducts such as formic acid, levulinic acid, and 5-HMF because of the severe reaction conditions at 190 °C (Figure 4).

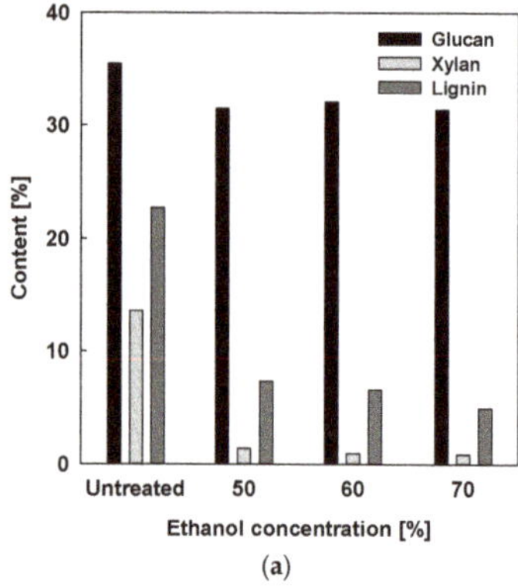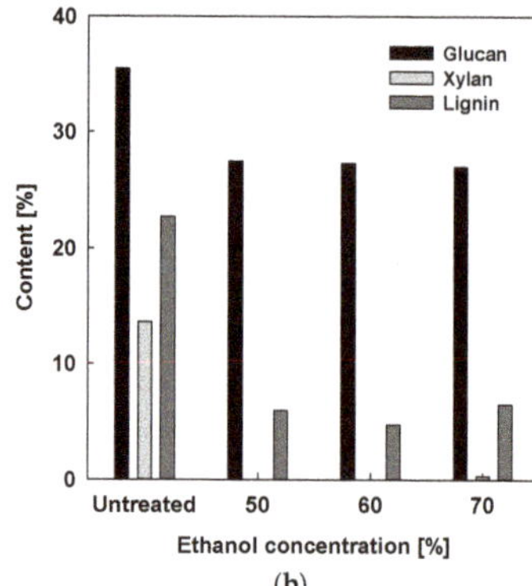

(a)   (b)

Figure 3. Effects of reaction temperature and EtOH concentration on the chemical compositions of raw and fractionated solid of rice husk: (**a**) 180 °C and (**b**) 190 °C. Note: reaction conditions: 180–190 °C, 50%–70% EtOH concentration, 0.25% (*w/v*) H_2SO_4 concentration.

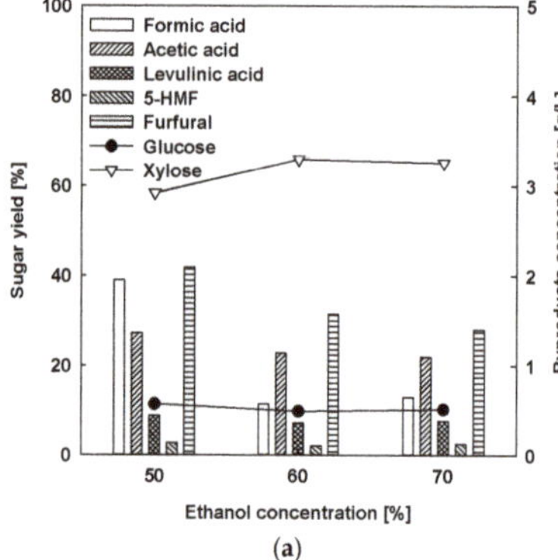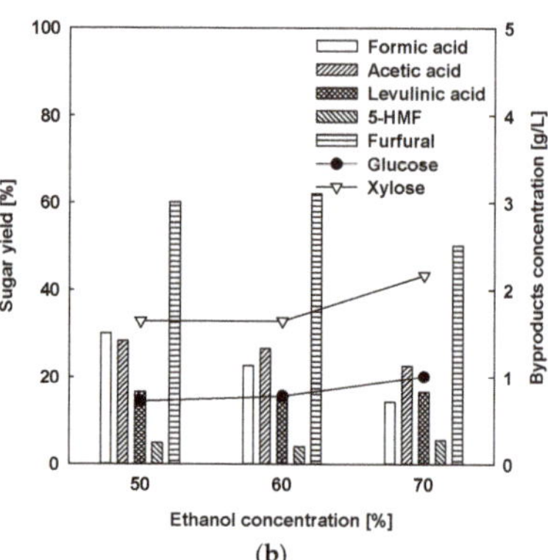

(a)   (b)

Figure 4. Effects of reaction temperature and EtOH concentration on the sugar extraction yield and the byproducts concentration of liquid hydrolyzate: (**a**) 180 °C and (**b**) 190 °C. Note: reaction conditions: 180–190 °C, 50%–70% EtOH concentration, 0.25% (*w/v*) H_2SO_4 concentration.

The delignification yield tended to increase with increasing EtOH concentration. In general, a high EtOH concentration increases the lignin solubility; however, Ni and Hu reported that lignin exhibited maximal solubility at an EtOH concentration of 70% [17]. If the EtOH concentration exceeds 70%, the lignin solubility decreases slightly. Therefore, the appropriate EtOH concentration for delignification is likely in the range from 60% to 70%.

2.1.2. Acid-Catalyzed Organosolv Fractionation for Xylose and Lignin Extraction

The effect of acid concentration in the organosolv fractionation is shown in Figure 5. As the SA concentration was increased from 0.15% to 0.7%, the glucan and xylan contents of the fractionated solid decreased (Figure 5a). At an SA concentration of 0.7%, xylan was not present and the amount of glucan was substantially decreased. The difference in SA concentration was confirmed more clearly in the hydrolyzate results in Figure 5b. The xylose extraction yield increased with increasing SA concentration from 0.15% to 0.25%. However, the highest xylose extraction yield (67.5%) was observed at 0.25% and tended to decrease with increasing SA concentration as the SA concentration was progressively increased to 0.7%. When hexose sugar (mainly glucose) is degraded under severe conditions, it is converted into formic acid, levulinic acid, and 5-HMF. A large amount of formic acid, levulinic acid, and 5-HMF were produced at SA concentrations of 0.35% and 0.7%, accompanied by a substantial decrease of the glucan content in the fractionated RH (Figure 5b). However, the change in lignin was inconsequential as the SA concentration was increased from 0.25% to 0.7%. This result means that an improvement of the delignification effect should not be expected at SA concentrations above a certain concentration. In the present system, the optimal SA concentration was determined to be 0.25%.

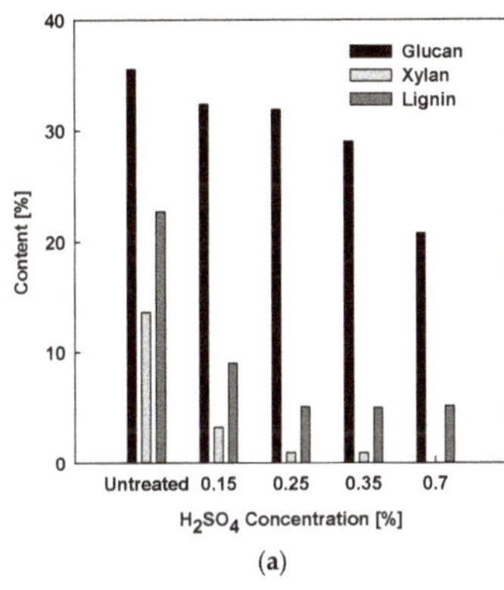
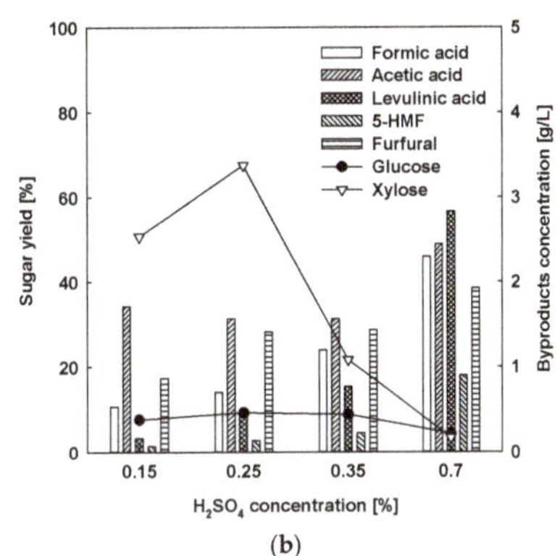

Figure 5. Effects of reaction temperature and H_2SO_4 concentration on (**a**) the chemical compositions of raw and fractionated solids of rice husk and (**b**) the sugar extraction yield and concentration of byproducts of liquid hydrolyzate. Note: reaction conditions: 180–190 °C, 60% EtOH concentration, 0.15%–0.7% (*w/v*) H_2SO_4 concentration.

2.1.3. Comparison with Acid-Free and Acid-Catalyzed Organosolv Fractionation

The chemical compositions of raw and fractionated RH and the extraction yields of glucose and xylose (under acid-free and acid-catalyzed conditions) are compared in Figure 6. The reaction temperature, reaction time, and EtOH concentration were fixed at 180 °C, 60 min, and 60% (*v/v*), respectively. As shown in Figure 6a, the glucan content did not show a dramatic difference under acid-free and acid-catalyzed conditions. By contrast, the xylose and lignin extraction yields were dramatically improved under acid catalysis. In addition, the xylose extraction yield under acid catalysis was also improved more than twofold (Figure 6b). Thus, in organosolv fractionation, SA can improve the xylose and lignin extraction yields.

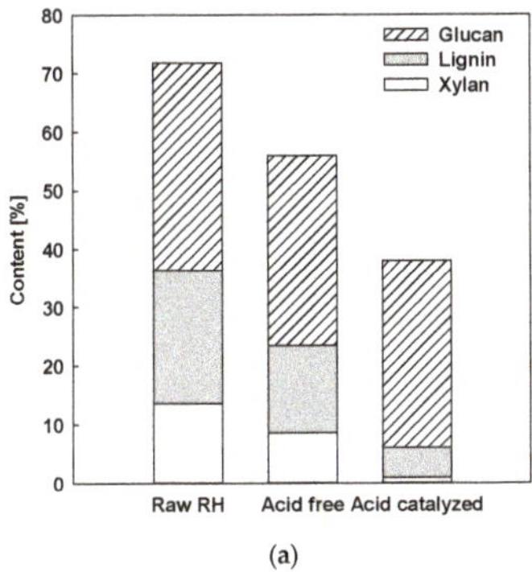
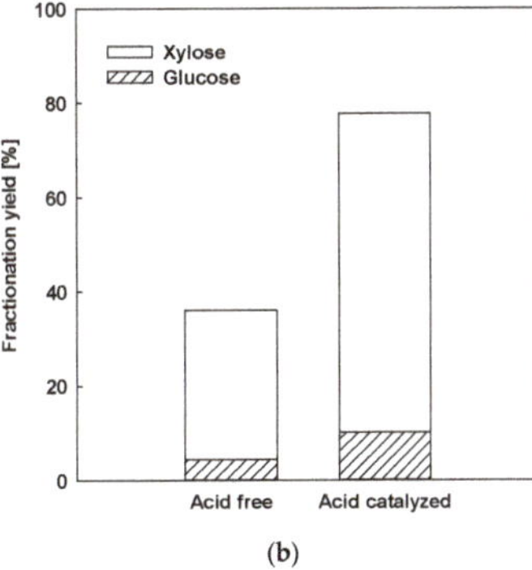

Figure 6. Comparison of acid-free and acid-catalyzed organosolv fractionations: (**a**) chemical composition of the fractionated RH and (**b**) the extraction yield of sugars. Note: reaction conditions: acid free: 180 °C, 60% EtOH concentration; acid catalyzed: 180 °C, 60% EtOH concentration, 0.25% H_2SO_4 (*w/v*) concentration.

2.1.4. Enzymatic Hydrolysis Tests

The enzymatic hydrolysis tests were performed with raw and fractionated RH; the results are present in Figure 7. The acid-catalyzed organosolv-fractionated RH showed 80.1% enzyme digestibility in 24 h, and the maximum digestibility was 85.2% at 72 h. By contrast, the acid-free organosolv-fractionated RH showed only 32.5% enzyme digestibility at 72 h and showed no appreciable difference in enzyme digestibility from the raw RH (24.5%). This lack of improvement in enzyme digestibility is attributed to xylan and phenolic compounds (lignin), which were insufficiently removed, interfering with cellulase access to the fractionated RH.

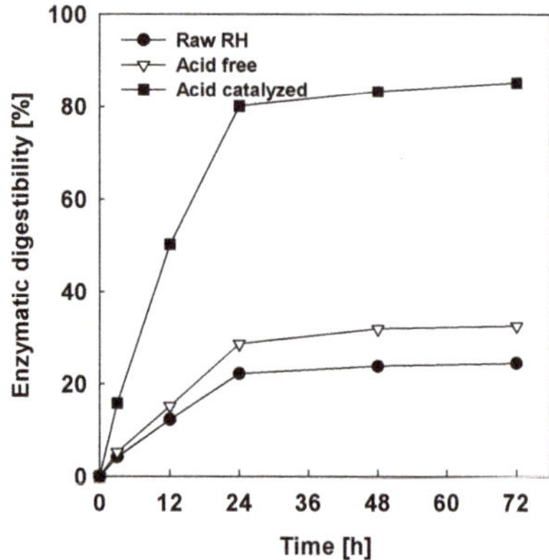

Figure 7. Enzymatic digestibility profiles of raw and fractionated RH; acid free: 180 °C, 60% EtOH concentration; acid catalyzed: 180 °C, 60% EtOH concentration, 0.25% H_2SO_4 (*w/v*) concentration. Note: enzymatic hydrolysis conditions: 15 FPU of Cellic® CTec2/g-glucan, pH 4.8, 50 °C, and 150 rpm.

2.2. Chemical Characteristics of Organosolv-Fractionated Lignin

The hydroxyl groups, which include aliphatic hydroxyl groups, phenolic hydroxyl groups (*p*-hydroxyphenyl units, guaiacyl units, and syringyl units), and carboxyl groups of lignin, were analyzed by ^{31}P NMR. The amounts of hydroxyl groups were calculated on the basis of an internal standard (cyclohexanol); the respective integrated peak areas are presented in Table 1. The M_n, M_w, and PD of lignin were determined by GPC; the results are presented in Table 1. As shown in Table 1, the acid-catalyzed organosolv-fractionated lignin contained higher concentrations of phenolic hydroxyl groups and carboxyl groups compared with the acid-free organosolv-fractionated lignin. This result is related to the severity of the reaction. In general, the β-aryl ether linkage is known to occupy 50% of the lignin structure [18]. With increasing reaction severity, phenolic hydroxyl and carboxyl groups generate more aromatic monomers through cleavage of the β-O-4 linkage [19]. Furthermore, cleaving β-O-4 linkages causes a decrease in lignin molecular weight [20]. These results are consistent with the molecular-weight results shown in Table 1, where the molecular weight of acid-catalyzed organosolv-fractionated lignin was higher than that of acid-free organosolv-fractionated lignin

Table 1. Hydroxyl groups and molecular weight of acid-free and acid-catalyzed organosolv precipitated lignin from RH.

Content	Classify	Unit	Acid-Free	Acid Catalyzed
Hydroxyl group	Aliphatic unit	mmol/g	2.84	2.76
	p-Hydroxyphenyl unit	mmol/g	0.49	0.50
	Guaiacyl unit	mmol/g	1.19	1.62
	Syringyl unit	mmol/g	0.54	0.56
	Phenols unit	mmol/g	2.21	2.68
	Carbonyl unit	mmol/g	0.04	0.09
Molecular weight	M_n [1]	g/mol	1296	1073
	M_w [2]	g/mol	1627	1422
	PDI [3]	-	1.26	1.33

[1] Number-average molecular weight, [2] Weight-average molecular weight, [3] Polydispersity index (M_w/M_n).

The previously presented results indicate that the acid-catalyzed organosolv-fractionated lignin has a low molecular weight, uniform molecular weight distribution, and a high concentration of phenolic hydroxyl groups. The phenolic hydroxyl groups increase the reactivity of lignin toward formaldehyde when aromatic polymers are used in phenolic resin formulations [21,22]. Therefore, lignin with these characteristics would be suitable for application in the biorefinery field.

2.3. Overall Fractionation Yield and Total Mass Balance

The extraction mass balance (EMB) of raw and fractionated RH was present in Table 2. In the acid-free organosolv fractionation, 91.8% of glucan was preserved from the fractionated RH and 31.6% of xylose and 11.0% of lignin were obtained from the liquid hydrolyzate. By contrast, in the acid-catalyzed organosolv fractionation, 89.9% of glucan was preserved from the fractionated RH and 67.6% of xylose and 49.8% of lignin were obtained from the liquid hydrolyzate. As the P lignin shown in Table 2 was lignin precipitated from liquid hydrolyzate, conclusively, three main components (glucan, xylose, and lignin) of RH were separated. Therefore, the acid-catalyzed organosolv pretreatment is a valuable process to fractionate three main components for RH in the biorefinery field.

Table 2. Extraction mass balance of sugars and lignin with acid-free acid acid catalyzed orgarnosolv fractionation of RH.

Sample		S.R. [%]	Solid [%]			Liquid [%]			EMB [1] [%]		
			Glucan	Xylan	Lignin	Glucose	Xylose	P. Lignin [2]	Glucan	Xylan	Lignin
Raw RH		100	35.5	13.6	22.7	-	-	-	-	-	-
Acid-Free	Fractionated	71.7	46.9	12.0	20.8	1.6	4.3	2.5	99.2	94.7	76.7
	Fractionated [3]		32.6	8.6	14.9						
	Component Retention [%]		91.8	63.7	65.6	-	-	-	-	-	-
Acid Catalyzed	Fractionated	50.9	62.7	2.0	10.0	3.3	9.2	11.3	99.2	75.6	72.2
	Fractionated [3]		31.9	1.0	5.1						
	Component Retention [%]		89.9	7.4	22.5	-	-	-	-	-	-

Fractionation conditions: Acid-free: 180 °C, 60 min, 60% (*v/v*) EtOH; acid catalyzed: 180 °C, 60 min, 60% (*v/v*) EtOH, 0.25% (*w/v*) H_2SO_4. [1] Extraction mass balance (EMB) = $(\sum C_{Li} + \sum C_{Si})/(\sum C_{Ri})$, where C_i is the mass of each component as C_{Li}, as determined through HPLC chromatography. The subscripts L, S, and R refer to the extracted liquid, fractionated solids, and raw fractions respectively. [2] Precipitated lignin from liquid hydrolyzate. [3] Data are based on the oven-dried raw biomass.

The simplified flowchart and an overall mass balance of the acid-catalyzed organosolv fractionation and consecutive enzymatic hydrolysis are summarized in Figure 8. Most of the hemicellulose and lignin were separated from the RH by organosolv fractionation with SA. Under the optimized conditions, approximately 49.9 g of the mass fraction was solubilized into liquid hydrolysate, which involved discharging 11.8 g of xylan, 17.6 g of lignin, and 4.7 g of glucan (based on 100 g of raw RH) from the reactor for recovery of lignin by the precipitation method. The residual solid, i.e., 50.1 g of acid-catalyzed fractionated RH, was subjected to consecutive enzymatic hydrolysis. The liquor resulting from the enzymatic hydrolysis had a sugar fraction including 30.0 g of glucose and 1.5 g of xylose, which could be easily used for microbial fermentation. The acid-catalyzed organosolv fractionation was assumed to have greatly increased the cellulase effectiveness through the removal of hemicellulose and lignin.

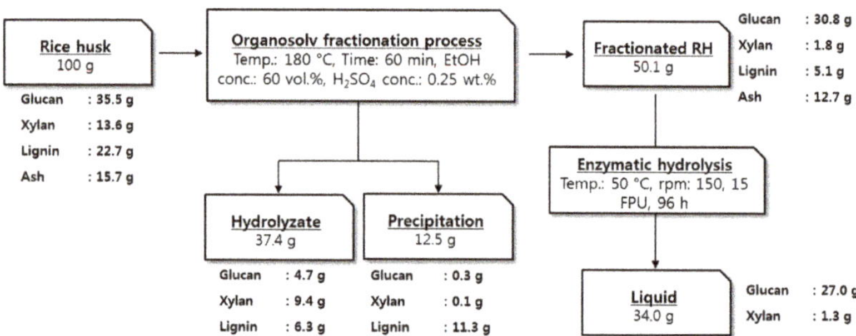

Figure 8. Overall mass balance of the organosolv fractionation process of RH under optimal conditions.

3. Materials and Methods

3.1. Materials

RH was harvested at Gimpo-si, Gyeonggi-do, Korea and collected in 2017. The RH was ground with a blend mill (Blender 7012s, Waring Commercial, Stamford, CT, USA) and then sieved to a nominal size of 14–45 mesh (from 0.36 to 1.4 mm). The ground RH was placed in a convection oven at 45 ± 5 °C for 48 h and then stored in an automatic dehumidification desiccator until used. The average moisture content of the dried RH was 4.3% during the experiment. The composition of the RH was determined by the National Renewable Energy Laboratory (NREL, Golden, CO, USA) Laboratory Standard Procedure (LAP) [23–28]. The chemical composition of the raw RH was 35.6% glucan, 13.6% xylan, 1.7% arabinan, 22.7% acid insoluble lignin (AIL), 0.7% acid-soluble lignin (ASL), 1.2% EtOH extractives, 6.6% water extractives, and 15.7% ash (n = 3, standard deviations < 0.8).

EtOH (cat. no. E7023), sodium azide (cat. no. S2002), tetrahydrofuran (THF, cat. no. 401757), pyridine (cat. no. 270970), chloroform-d (cat. no. 151858), cyclohexanol (cat. no. 105899), chromium(III) acetylacetonate (cat. no. 574082), 2-chloro-4,4,5,5-tetramethyl-1,2,3-dioxaphospholane (TMDP, cat. no. 447536), and cellulase enzyme Cellic® CTec2 (Novozymes, A/S Bagsvaerd, Denmark) were purchased from Sigma-Aldrich Korea.

3.2. Experimental Setup and Operation

The batch reactor used for the organosolv fractionation process consisted of reaction baths (a molten salt bath and a silicone oil bath) and a cooling bath (water bath). The temperatures of the molten salt bath and the silicone oil bath could be driven to 250 °C and 200 °C, respectively. The molten salt bath was used for preheating to the target temperature, the silicone oil bath was used for maintaining the reaction temperature, and the water bath was used for cooling. The average preheating time was less than 1.0 min under all of the investigated reaction temperature conditions. The fractionation reactor, a bomb tubular reactor, was constructed of SS-316L tubing with a 10.9 mm ID and a 150 mm length (14.0 cm³ internal volume). The temperatures of the reaction baths and fractionation reactors were measured continuously with high-temperature thermocouples (catalog number HY-72D, Hanyoung Nux, Incheon, Korea). The timer and movement controller were set up to control the reaction time and movement of the fractionation reactors, respectively.

When the reaction was completed, the liquid samples were removed from the reactor, diluted threefold with deionized (DI) water to precipitate the lignin, and then evaporated in a drying oven at 55 °C for 4 h. The liquid samples were analyzed to determine their concentrations of carbohydrates (i.e., glucose, xylose, and arabinose) and byproducts. The fractionated solids discharged from the reactor were separated into two portions. One portion was dried using a convection oven for weight-loss measurement and composition analysis. The other portion was subjected to an enzymatic digestibility test in the wet state.

3.3. Enzymatic Digestibility Tests

The enzymatic digestibility test of raw and fractionated RH was determined according to the NREL LAP [27]. The tests were conducted under the following conditions: 50 °C, pH 4.8 (0.05 M sodium citrate buffer), 150 rpm in a shaking incubator (model VS-8480SFN, Vision Scientific Co., Bucheon, Korea) and 15 FPU g-glucan enzyme loadings. The average activity of the cellulase was measured to be 119.4 FPU/mL. The initial glucan concentration was 1.0% (w/v) based on 100 mL of total liquid in a 250 mL Erlenmeyer flask. To prevent microbial contamination, 1.0 mL of 20 mg/mL sodium azide was added. Samples were collected periodically at appropriate sampling times (6, 12, 24, 48, and 72 h) and analyzed for hydrolyzed glucose using a high-performance liquid chromatography (HPLC) system.

3.4. Composition Analysis of Raw and Fractionated RH

The chemical compositions of the solid and liquid samples were determined according to the procedures of the NREL-LAP [24,25,28]. The extractives process was carried out in two steps using water and EtOH consecutively. For the composition analysis of extractives-free and fractionated solids, two-step acid hydrolysis was carried out.

An HPLC system (LC-10A, Shimadzu Inc., Kyoto, Japan) with a refractive index (RI) detector (RID-10A, Shimadzu Inc., Kyoto, Japan) was used to determine the carbohydrate and organic acid components of the samples. For analysis of monomeric sugars from the raw and fractionated RH samples, a carbohydrate column (Aminex HPX-87P, Bio-Rad Inc., Hercules, CA, USA) was used; HPLC-grade water was used as the mobile phase with a volumetric flow rate of 0.4 mL/min. The samples were neutralized with calcium carbonate and filtered (0.2 μm pore size) before analysis. The operating temperature of the column was 80 °C. The liquid hydrolysis samples and enzymatic hydrolysis samples were analyzed using an organic acid column (Aminex HPX-87H, Bio-Rad Inc., Hercules, CA, USA); 5 mM SA was used as a mobile phase with a volumetric flow rate of 0.5 mL/min. These samples were also neutralized with calcium carbonate and filtered (0.4 μm pore size) before analysis. The operating temperature of the column was 65 °C.

3.5. Chemical Characterization of Organosolv Fractionated Lignin

The number-average molar mass (M_n), weight-average molar mass (M_w), and polydispersity (PD) of the organosolv fractionated lignin samples were determined by gel permeation chromatography (GPC, Ultimate 3000, Thermo Fisher Scientific Inc., Waltham, MA, USA). For molecular-weight determination, 3 mg of an acetylated lignin sample was dissolved in 2 mL of THF and filtered with a 0.45 μm polytetrafluoroethylene (PTFE) syringe filter to remove impurities. The GPC system was equipped with a Shodex column (KF-806L with Shodex KF-G guard column, Showa Denko, Tokyo, Japan) and an RI detector (Refracto Max 520, Thermofisher Scientific Korea Ltd, Seoul, Korea); THF was used as the mobile phase (1.0 mL/min); the injection volume was 20 μL.

^{31}P NMR (Avance 600, Bruker, Billerica, MA, USA) spectra were recorded at 242.88 MHz and 256 scans with a 2 s pulse delay. For quantitative ^{31}P NMR analysis, 20 mg of lignin sample was accurately weighed and dissolved in 400 μL of solution A and 150 μL of solution B in a 5 mL vial. Solution A was a mixture of pyridine and chloroform-d ($CDCl_3$) at a ratio of 1.6:1 (v/v). Solution B was a mixture of solution A (25 mL), cyclohexanol (100 mg), and chromium(III) acetylacetonate (90 mg). The dissolved liquid was vortexed for 5 min, 70 μL of TMDP was added to the solution, and the resultant mixture was analyzed with the ^{31}P NMR system.

Fourier transform infrared (FTIR) spectroscopy (IRSpirit-L/T, Shimadzu Inc., Kyoto, Japan) was used to determine the characteristic absorption peaks of the chemical functional groups in the organosolv-fractionated lignin via the attenuated total reflectance (ATR) technique. Mid-IR spectra were collected by averaging 40 scans collected at a resolution of 1 cm^{-1} over the wavenumber region from 4000 to 500 cm^{-1}.

4. Conclusions

The results of this work indicate that acid-catalyzed organosolv fractionation using a low reaction severity (i.e., low acid concentration) is a worthwhile process in the field of biorefinery. Very pure lignin fractions were recovered, and the resulting hydrolyzate, xylose-rich liquid fraction can potentially be used for xylose-uptaking fermentations; similarly, the insoluble residue, the cellulose-rich solid fraction, could potentially be readily hydrolyzed at a low enzyme loading into glucose. These findings strongly suggest that a biorefinery procedure is required prior to a five-carbon fermentation process, as well as catalytic conversion of fractionated lignin, for full valorization of the cellulosic biomass.

Author Contributions: T.H.K., H.J.R., and K.K.O. contributed equally to this work. T.H.K. contributed to the experimental process for RH Fractionation and T.H.K. contributed to the project administration and experimental design. And H.J.R. contributed to providing the methodology and data validation. All of the authors contributed to the writing and review of this document.

Funding: This work was supported by the R and D program of Korea Institute of Energy Technology Evaluation and Planning (KETEP) grant funded by the Ministry of Trade, Industry and Energy (MOTIE), the Republic of Korea (No. 20183030091950).

Conflicts of Interest: The authors declare no conflict of interest.

References

1. Lim, J.S.; Manan, Z.A.; Alwi, S.R.W.; Hashim, H. A review on utilisation of biomass from rice industry as a source of renewable energy. *Renew. Sustain. Energy Rev.* **2012**, *16*, 3084–3094. [CrossRef]
2. Zhang, H.; Ding, X.; Chen, X.; Ma, Y.; Wang, Z.; Zhao, X. A new method of utilizing rice husk: Consecutively preparing d-xylose, organosolv lignin, ethanol and amorphous superfine silica. *J. Hazard. Mater.* **2015**, *291*, 65–73. [CrossRef]
3. Soltani, N.; Bahrami, A.; Pech-Canul, M.I.; González, L.A. Review on the physicochemical treatments of rice husk for production of advanced materials. *Chem. Eng. J.* **2015**, *264*, 899–935. [CrossRef]
4. Menya, E.; Olupot, P.W.; Storz, H.; Lubwama, M.; Kiros, Y. Production and performance of activated carbon from rice husks for removal of natural organic matter from water: A review. *Chem. Eng. Res. Des.* **2018**, *129*, 271–296. [CrossRef]
5. Kim, T.H.; Kim, T.H. Consecutive Recovery of Non-Structural Sugars and Xylooligomers from Corn Stover using Hot Water and Acidified Calcium Chloride. *BioResources* **2018**, *13*, 7294–7309. [CrossRef]
6. International Energy Agency. Available online: https://www.ieabioenergy.com/publications/bio-based-chemicals-value-added-products-from-biorefineries/ (accessed on 14 March 2019).
7. Chen, H.; Liu, J.; Chang, X.; Chen, D.; Xue, Y.; Liu, P.; Lin, H.; Han, S. A review on the pretreatment of lignocellulose for high-value chemicals. *Fuel Process. Technol.* **2017**, *160*, 196–206. [CrossRef]
8. Guo, Y.; Zhou, J.; Wen, J.; Sun, G.; Sun, Y. Structural transformations of triploid of Populus tomentosa Carr. lignin during auto-catalyzed ethanol organosolv pretreatment. *Ind. Crops Prod.* **2015**, *76*, 522–529. [CrossRef]
9. Zhao, X.; Cheng, K.; Liu, D. Organosolv pretreatment of lignocellulosic biomass for enzymatic hydrolysis. *Appl. Microbiol. Biotechnol.* **2009**, *82*, 815. [CrossRef] [PubMed]
10. Zhang, K.; Pei, Z.; Wang, D. Organic solvent pretreatment of lignocellulosic biomass for biofuels and biochemicals: A review. *Bioresour. Technol.* **2016**, *199*, 21–33. [CrossRef]
11. Lora, J.H.; Glasser, W.G. Recent industrial applications of lignin: A sustainable alternative to nonrenewable materials. *J. Polym. Environ.* **2002**, *10*, 39–48. [CrossRef]

12. Teramoto, Y.; Tanaka, N.; Lee, S.H.; Endo, T. Pretreatment of eucalyptus wood chips for enzymatic saccharification using combined sulfuric acid-free ethanol cooking and ball milling. *Biotechnol. Bioeng.* **2008**, *99*, 75–85. [CrossRef] [PubMed]

13. Kim, T.H.; Im, D.J.; Oh, K.K. Effects of Organosolv Pretreatment Using Temperature-Controlled Bench-Scale Ball Milling on Enzymatic Saccharification of Miscanthus × giganteus. *Energies* **2018**, *11*, 2657. [CrossRef]

14. Pan, X.; Kadla, J.F.; Ehara, K.; Gilkes, N.; Saddler, J.N. Organosolv ethanol lignin from hybrid poplar as a radical scavenger: Relationship between lignin structure, extraction conditions, and antioxidant activity. *J. Agric. Food Chem.* **2006**, *54*, 5806–5813. [CrossRef] [PubMed]

15. Weinwurm, F.; Turk, T.; Denner, J.; Whitmore, K.; Friedl, A. Combined liquid hot water and ethanol organosolv treatment of wheat straw for extraction and reaction modeling. *J. Clean. Prod.* **2017**, *165*, 1473–1484. [CrossRef]

16. Kim, T.H.; Jeon, Y.J.; Oh, K.K.; Kim, T.H. Production of furfural and cellulose from barley straw using acidified zinc chloride. *Korean J. Chem. Eng.* **2013**, *30*, 1339–1346. [CrossRef]

17. Pan, X.; Gilkes, N.; Kadla, J.; Pye, K.; Saka, S.; Gregg, D.; Ehara, K.; Xie, D.; Lam, D.; Saddler, J. Bioconversion of hybrid poplar to ethanol and co-products using an organosolv fractionation process: Optimization of process yields. *Biotechnol. Bioeng.* **2006**, *94*, 851–861. [CrossRef]

18. Agarwal, A.; Rana, M.; Park, J.H. Advancement in technologies for the depolymerization of lignin. *Fuel Process. Technol.* **2018**, *181*, 115–132. [CrossRef]

19. Yáñez-S, M.; Matsuhiro, B.; Nunez, C.; Pan, S.; Hubbell, C.A.; Sannigrahi, P.; Ragauskas, A.J. Physicochemical characterization of ethanol organosolv lignin (EOL) from Eucalyptus globulus: Effect of extraction conditions on the molecular structure. *Polym. Degrad. Stab.* **2014**, *110*, 184–194. [CrossRef]

20. Domínguez-Robles, J.; Tamminen, T.; Liitiä, T.; Peresin, M.S.; Rodríguez, A.; Jääskeläinen, A.S. Aqueous acetone fractionation of kraft, organosolv and soda lignins. *Int. J. Biol. Macromol.* **2018**, *106*, 979–987. [CrossRef]

21. El Mansouri, N.E.; Salvadó, J. Structural characterization of technical lignins for the production of adhesives: Application to lignosulfonate, kraft, soda-anthraquinone, organosolv and ethanol process lignins. *Ind. Crops Prod.* **2006**, *24*, 8–16. [CrossRef]

22. Fahmi, R.; Bridgwater, A.V.; Donnison, I.; Yates, N.; Jones, J.M. The effect of lignin and inorganic species in biomass on pyrolysis oil yields, quality and stability. *Fuel* **2008**, *87*, 1230–1240. [CrossRef]

23. Hames, B.; Ruiz, R.; Scarlata, C.; Sluiter, A.; Sluiter, J.; Tmpleton, D. *Preparation of Samples for Compositional Analysis*; NREL/TP-510-42620; National Renewable Energy Laboratory: Golden, CO, USA, 2012.

24. Sluiter, A.; Ruiz, R.; Scarlata, C.; Sluiter, J.; Templeton, D. *Determination of Extractives in Biomass*; NREL/TP-510-42619; National Renewable Energy Laboratory: Golden, CO, USA, 2012.

25. Sluiter, A.; Hames, B.; Ruiz, R.; Scarlata, C.; Sluiter, J.; Templeton, D. *Determination of Structural Carbohydrates and Lignin in Biomass*; NREL/TP-510-42618; National Renewable Energy Laboratory: Golden, CO, USA, 2012.

26. Sluiter, A.; Hames, B.; Ruiz, R.; Scarlata, C.; Sluiter, J.; Templeton, D. *Determination of Ash in Biomass*; NREL/TP-510-42622; National Renewable Energy Laboratory: Golden, CO, USA, 2008.

27. Sluiter, A.; Hames, B.; Ruiz, R.; Scarlata, C.; Sluiter, J.; Templeton, D. *Determination of Sugars, Byproducts, and Degradation Products in Liquid Fraction Process Samples*; NREL/TP-510-42623; National Renewable Energy Laboratory: Golden, CO, USA, 2008.

28. Selig, M.; Weiss, N.; Ji, Y. *Enzymatic Saccharification of Lignocellulosic Biomass*; NREL/TP-510-42629; National Renewable Energy Laboratory: Golden, CO, USA, 2008.

Article

Kinetic Characterization of Enzymatic Hydrolysis of Apple Pomace as Feedstock for a Sugar-Based Biorefinery

Alessandra Procentese [1], Maria Elena Russo [1,*], Ilaria Di Somma [1] and Antonio Marzocchella [2]

[1]   Istituto di Ricerche sulla Combustione, Consiglio Nazionale delle Ricerche, 80125 Napoli, Italy; alessandra.procentese@irc.cnr.it (A.P.); ilaria.disomma@irc.cnr.it (I.D.S.)

[2]   Dipartimento di Ingegneria Chimica, dei Materiali e della Produzione Industriale, Università degli Studi di Napoli Federico II, 80125 Napoli, Italy; antonio.marzocchella@unina.it

*   Correspondence: address: mariaelena.russo@cnr.it; Tel.: +39-0817682969

Received: 4 February 2020; Accepted: 24 February 2020; Published: 26 February 2020

Abstract: The enzymatic hydrolysis of cellulose from biomass feedstock in the sugar-based biorefinery chain is penalized by enzyme cost and difficulty to approach the theoretical maximum cellulose conversion degree. As a consequence, the process is currently investigated to identify the best operating conditions with reference to each biomass feedstock. The present work reports an investigation regarding the enzymatic hydrolysis of apple pomace (AP). AP is an agro-food waste largely available in Europe that might be exploited as a sugar source for biorefinery purposes. A biomass pre-treatment step was required before the enzymatic hydrolysis to make available polysaccharides chains to the biocatalyst. The AP samples were pre-treated through alkaline (NaOH), acid (HCl), and enzymatic (laccase) delignification processes to investigate the effect of lignin content and polysaccharides composition on enzymatic hydrolysis. Enzymatic hydrolysis tests were carried out using a commercial cocktail (Cellic®CTec2) of cellulolytic enzymes. The effect of mixing speed and biomass concentration on the experimental overall glucose production rate was assessed. The characterization of the glucose production rate by the assessment of pseudo-homogeneous kinetic models was proposed. Data were analysed to assess kinetic parameters of pseudo-mechanistic models able to describe the glucose production rate during AP enzymatic hydrolysis. In particular, pseudo-homogeneous Michaelis and Menten, as well as Chrastil's models were used. The effect of lignin content on the enzymatic hydrolysis rate was evaluated. Chrastil's model provided the best description of the glucose production rate.

Keywords: enzymatic hydrolysis; waste biomass; kinetics; biomass pre-treatment

---

1. Introduction

The conversion of lignocellulosic biomass into bio-based chemicals according to the biorefinery sugar-platform involves four main steps: (i) pre-treatment to remove/disaggregate the lignin structure; (ii) enzymatic hydrolysis to convert polysaccharides into fermentable sugars; (iii) fermentation to convert sugars into bio-based products; and (iv) recovery and concentration of the bio-based products [1]. The enzymatic conversion of cellulose into fermentable sugars is one of the bottlenecks of the whole biorefinery chain. Indeed, the success of this step depends strongly on the operating conditions (e.g., temperature, pH, biomass/liquid ratio, enzyme loading), the effect of product inhibition on enzyme catalysis [2], and the enzyme consumption (specific activity, stability, enzyme recycling strategies). Altogether, these issues affect the overall production cost of the sugars. Zhang et al. [3] pointed out that the incidence of the pre-treatment cost on the biofuel production is about 0.13–0.17 € per litre of ethanol, a quite large fraction for the competitive biofuel cost that should be less than 0.6 €/L. In addition, the first pre-treatment aimed at delignification strongly affects the rate and the degree of conversion of the enzymatic hydrolysis of polysaccharides. In particular, the nature of the pre-treatment (steam explosion, chemical pre-treatment with acid or alkaline solvents, organosolv, biological, etc.) influence the composition and the structure of the biomass substrate subjected to the enzymatic hydrolysis [4,5].

Some technical issues related to the enzymatic hydrolysis process are still open. Among these issues, the selection of the reactor configuration and the optimization of sugar recovery can be tackled based on quantitative information about enzymatic hydrolysis kinetics. With regard to the reactor configuration, different valuable studies date back to the eighties [6,7]. Tan et al. [6] carried out the enzymatic hydrolysis in a column reactor with continuous cellulase recycling to decrease the enzyme consumption. Gusakov et al. [7] tested the cellulose conversion in a batch reactor and in a plug-flow reactor to select the best configuration in terms of cellulose conversion. Nowadays, further studies are focused on innovative strategies to combine two fundamental targets: the maximization of both sugar concentration in the hydrolysate and polysaccharides conversion [8,9]. Typically, these two targets are in contrast with one another: high biomass loading (15–20 wt%) favours high soluble sugar concentration and low polysaccharides conversion, and low biomass loading (about 5 wt%) provides low sugar concentration and high polysaccharides conversion. An effective strategy was proposed by Xu et al. [8] to overcome this issue. They reported a two-step process that provided cellulose conversion and a sugar concentration in the final hydrolysate that is higher than the conventional single-step hydrolysis. The two-step process was based on the optimization of the enzyme adsorption on the biomass during the first short stage of the process (about 1 h) and the successive optimization of extensive hydrolysis during the second, long stage of the process (up to 48 h). A similar strategy was modelled by González Quiroga et al. [9]. They proposed a two-step enzymatic hydrolysis at different biomass loadings in different continuous reactors.

The further development of the proposed two-step strategies asks for the quantitative characterization of the hydrolysis of cellulose in real biomass. Hydrolysis is a heterogeneous process that includes [10,11] liquid–solid transfer of biocatalyst, enzymes adsorption on biomass surface, enzyme complexation with the polysaccharide chains, and cellulose and hemicellulose enzymatic hydrolysis. The last step may be affected by product inhibition, too. The rate of each of the listed phenomena depends on biomass structure, biomass loading, type of biocatalyst, and reactor configuration. As a consequence, the overall sugar production rate may be a complex combination of each step and its characterization can provide a rational tool for process design. Many kinetic models have been applied to the enzymatic hydrolysis of cellulose [9,10]. However, to the author's knowledge, only some studies refer to the kinetic characterization of the enzymatic hydrolysis of real biomasses after delignification pre-treatment [12–14]. These studies provide kinetic parameters to be used to identify the optimal enzymatic hydrolysis operating conditions and to help to quantify the effect of the biomass composition (sugar and lignin content) resulting from different delignification conditions.

The present work proposes the kinetic characterization of the enzymatic hydrolysis of real biomass according to rigorously controlled conditions. Residues from apple pomace (AP) processing were used as substrate because it is one of the most available European agro-food wastes [15]. The work was aimed at investigating the effect of mixing rate applied to the biomass slurry during the enzymatic hydrolysis; quantifying through kinetic parameters the dependence of glucose production rate on the biomass composition affected in its turn by the delignification process, selecting a proper pseudo-homogeneous kinetic model for the description of the process. To these aims, AP samples were pre-treated with a chemical (alkaline or acid) and biological (laccases) pre-treatment and hydrolysed using commercial enzymes as biocatalysts, and then enzymatic hydrolysis tests were performed under a relevant range of mixing speeds, enzyme concentrations, and biomass-to-liquid ratios. Pseudo-homogeneous models were adopted for the kinetic characterization, and the kinetic parameters of the Michaelis and Menten with product inhibition model and that of Chrastil's model were assessed.

2. Materials and Methods

2.1. Raw Material and Eenzymes

Apple pomace (AP) was selected as the waste biomass in the framework of the European project "Waste2Fuel" (grant agreement N°654623) and was kindly supplied by Muns Agroindustrial S.L. (Lleida, Spain). The raw biomass samples were dried at 40 °C, sieved in the range 0.5–1 mm, and stored in sealed plastic bags at room temperature until used.

The enzyme cocktail Cellic®CTec2 was selected according to the common procedure followed in the framework of the European project "Waste2Fuel" (grant agreement N°654623). Moreover, the use of Cellic®CTec2 allowed to compare the data with those reported by Pratto et al. [13]. The commercial cocktail Cellic®CTec2 was kindly donated by Novozymes Latin America (Araucária, PR, Brazil); it has an undisclosed composition and is properly made by a mixture of cellulases, hemicellulases, and β-glucosidases so that pentose and hexose sugars can be recovered as products of polysaccharides hydrolysis. The cellulase activity was determined in terms of Filter Paper Units (FPU) according to Adney and Baker [16].

2.2. Biomass Pretreatment

Any saccharification process of lignocellulose biomass needs removal of lignin to make the polysaccharides chains accessible to the enzymes in the enzymatic hydrolysis step. Several chemical and biological delignification pre-treatments have been proposed as alternative to the conventional steam explosion process, to minimize energy duty and the formation of inhibitors compound affecting the fermentation of hydrolytic sugars. In the present study, AP was pre-treated according to three processes: hydrolysis in diluted alkaline solvent, hydrolysis in diluted acid solvent, and laccase catalysed delignification. Details on pre-treatment procedures are reported in the next subsections.

2.2.1. Chemical Delignification

Chemical delignification was accomplished by using diluted alkaline or acid solvents. Dried AP powder was soaked in 2% wt NaOH or HCl aqueous solution: 10 mL of solution were used per gram of solid mass. The suspension was kept in an autoclave (VAPORMATIC 770) for 30 min at 121 °C according to the procedure suggested by Procentese et al. (2017). The biomass was separated from the liquid phase (black liquor) by 10 min centrifugation at 5000 rpm. The recovered biomass was washed with distilled water until pH 7 was reached, then it was dried at 40 °C, weighted, and stored until used for enzymatic hydrolysis.

2.2.2. Laccase Catalysed Delignification

Dried AP powder was treated with laccases (166 U mL^{-1}), kindly provided by Biopox s.r.l., in a 50 mM sodium citrate buffer of pH 5. Tests were carried out by setting the dry biomass to solvent ratio at 1:10, in a shaker incubator at 150 rpm and 28 °C for 24 h. After pre-treatment, the biomass was filtered and washed with water, dried at 38 °C until a constant weight was reached, and then weighted and stored until used for enzymatic hydrolysis.

2.3. Enzymatic Hydrolysis of Polysaccharides

Enzymatic hydrolysis tests were carried out using the abovementioned commercial cocktail Cellic®Ctec2. The optimal temperature and pH for Cellic®CTec2 are 50 °C and 4.8, respectively. The activity of the sample cocktail, used for the present investigation, was about 142 FPU mL^{-1}. Enzymatic hydrolysis of pre-treated AP samples was carried out in a 250 mL overhead stirred batch reactor (Applikon MiniBio Lab Reactor Bundles) at a controlled stirring speed and temperature (50 °C) with a 100 mL reaction volume in a 0.05 M sodium citrate buffer of pH 4.8; all the tests were carried out in triplicate. Two hydrolysis tests were performed for each biomass type: short-term tests (1 h) and long-term tests (72 h). The short-term tests were carried out to assess the effects of mixing rate and of substrate concentration on the glucose production rate. The long-term tests were carried out to assess the effect of product as inhibition of the glucose production rate.

2.4. Effect of Mixing Speed

Short-term hydrolysis tests of pre-treated AP samples were carried out to assess the initial glucose production rate according to Pratto et al. [4], setting the mixing rate between 50 and 500 rpm. Substrate and enzyme concentrations were set at 10% ($m_{solid}/v_{solution}$) and 5 FPU $g_{cellulose}^{-1}$, respectively [4]. Set the mixing speed, the glucose production rate was assessed as the amount of produced glucose after 1 h. The amount of produced glucose was calculated as the difference between the initial glucan concentration (mass of glucan in the biomass per unit volume of liquid) and the glucose concentration in the liquid.

The optimal mixing speed was selected as the minimum value above which no further increase in glucose production rate was observed. The optimal value of the mixing speed was set in tests focused on the investigation of the effects of other operating conditions.

2.5. Effect of Substrate Concentration

The effect of substrate concentration was assessed through brief-term enzymatic hydrolysis tests by changing the biomass loading in the range 1–12% w/v. The enzyme concentration was fixed at 290 FPU $L_{solution}^{-1}$ and the stirring speed was fixed at the optimal value selected according to the tests described in Section 2.6. The glucose production rate was measured, at each biomass concentration, as the amount of produced glucose after 1 h.

2.6. Effect of Product Inhibition

Product inhibition is well documented in the literature on kinetics of cellulase enzymes [4,7,12,14]. The use of commercial cocktails secures the formation of monomeric sugars; thus, glucose is the main product of polysaccharides hydrolysis. Long-term tests were carried out to evaluate the effect of glucose inhibition on the overall enzymatic rate of glucose production. The tests were carried out at a 5%, 7.5%, and 10% w/v biomass loading for 72 h. The enzyme concentration was set at 828 FPU L^{-1}.

2.7. Analytical Methods

The raw and pre-treated biomass samples were characterized in terms of glucan, xylan, arabinan, and lignin content according to Sluiter et al. [17] by a two-step acid hydrolysis. About 0.3 g biomass sample was suspended in 72% H_2SO_4 and kept under mixing at 30 °C for 60 min. Then, the acid solvent was diluted to 4% H_2SO_4 by adding distilled water; the suspension was kept in an autoclave at 121 °C for 60 min. After sample cooling, the solid fraction was vacuum filtered and repeatedly washed with distilled water. The liquid fraction was recovered, neutralized to pH 5–6 by $CaCO_3$ addition, and analysed for monomeric sugars concentrations through an HPLC (Agilent 1260 Infinity HPLC system equipped with a refractive index detector and a Rezex RHM-Monosaccharide H+ Phenomenex column (300 × 7.8 mm, 8 μm), City, state abbreviation if USA, Country) and for acid soluble lignin through UV spectroscopy (City, state abbreviation if USA, Country) [17]. The solid fraction was dried at 105 °C until a constant weight was achieved and then weighted for the assessment of the acid insoluble residue.

The glucose concentration, in samples withdrawn during hydrolysis tests, was assessed after biomass separation (centrifugation at 5000 rpm for 10 min) and liquid filtration (0.25 μm cut-off). The enzymatic kit "D-glucose HK" (Megazyme) was used to assess the glucose concentration in the liquid samples.

2.8. Data Analysis

The experimental data from the hydrolysis tests were worked out to assess the overall kinetics of glucose production during the enzymatic hydrolysis of the pre-treated AP. The presence of oligosaccharides as an intermediate product of hydrolysis (e.g., cellobiose) was neglected and glucose was considered as the only soluble product because the commercial cellulase cocktail Cellic®CTec2 is characterized by a large content of β-glucosidase. The semi-mechanistic kinetic models reported hereinafter have been adopted.

Michaelis and Menten model. The cellulose hydrolysis in AP can be described by the Michaelis and Menten model (Equation (1)), assuming that (1) the substrate is a soluble reactant; (2) the system is homogeneous; (3) the concentration of the enzyme is constant; (4) the formation of the enzyme-substrate complex is rapid and reversible; (5) the breakdown of the enzyme-substrate complex into products is the limiting step of the overall reaction; and (6) the formation of the products is irreversible. The Michaelis and Menten model can be used to determine the initial reaction rate as well as how the glucose production rate proceeds along time.

$$V_{(S)} = \frac{V_{Max}\, S}{K_M + S} \tag{1}$$

where V_{Max} ($= k^*E_0$) is the maximum rate of reaction at fixed enzyme concentration (E_0), S the cellulose concentration, and K_M the Michaelis and Menten constant. The Michaelis and Menten model has also been applied in the form reported in Equation (2) to take into account the competitive enzyme inhibition by glucose.

$$V_{(S)} = \frac{V_{Max}\, S}{K_M\left(1 + \frac{P}{K_i}\right) + S} \tag{2}$$

where P is the glucose concentration, and K_i the product inhibition constant. The Michaelis and Menten model with inhibition by glucose was considered to describe the overall glucose production rate, resulting from long term hydrolysis tests.

Chrastil's model. This model was developed to describe kinetics taking into account the structural characteristics of a heterogeneous system [18]. According to this model, the time course of glucose concentration can be described by Equation (3):

$$P = P_\infty \left[1 - \exp(-kE_0 t)\right]^n \tag{3}$$

where P and P_∞ are the product concentration at time t and at the equilibrium (maximum conversion degree), respectively, k is the rate constant proportional to the diffusion coefficient, E_0 the initial enzyme concentration, and n a diffusion resistance constant that depends on the structure of the substrate and indicates the apparent reaction order. For negligible diffusion resistance, n tends to 1 and for limiting diffusion resistances n < 1.

Rate of glucose production vs. substrate concentration from data of short- and long-term tests were regressed according to the procedure described in the "Supplementary Materials" section.

3. Results and Discussion

3.1. Characterization of Raw and Pretreated Biomass

Raw and pre-treated AP samples were characterized in term of cellulose, hemicellulose, and lignin content according to the polysaccharides quantification assay [17] as well as in terms of biomass recovery after each delignification pre-treatment. Results of the biomass recovery and characterization are reported in Table 1. The highest biomass recovery was found for AP samples pre-treated with laccase (75%); the smallest biomass recovery was found after acid hydrolysis of AP (50%). The highest glucan content was measured for biomass samples pre-treated with NaOH; the smallest glucan content was measured for AP samples pre-treated with laccases. AP pre-treated with laccases retained the highest lignin fraction among the pre-treated AP samples (14.5% AIL). The degree of delignification was equal to 12%, 70%, and 45% for biomass pre-treated with laccase, NaOH, and HCl, respectively.

Table 1. Characterization of raw and pre-treated apple pomace (AP) in terms of polysaccharides and lignin content (Acid Insoluble Lignin—AIL, Acid Soluble Lignin—ASL).

Pre-Treatment	Recovery (%)	Composition (%)			
		Glucans	Xylans and Arabinans	AIL	ASL
Raw	100	21.2 ± 0.01	14.75 ± 0.03	16.5 ± 0.55	2.1 ± 0.34
NaOH	60	28.0 ± 0.01	0.8 ± 0.05	5.0 ± 0.53	n.d
HCL	50	25.0 ± 0.01	0.2 ± 0.06	9.0 ± 0.43	n.d
Laccases	75	22.5 ± 0.01	12.04 ± 0.03	14.5 ± 0.54	n.d

The observed different performances in terms of pre-treated biomass composition are due to the type of pre-treatment; in fact, two of the three adopted pre-treatments act with an unspecific hydrolysis by an aqueous alkaline or acid solvent, and the biological pre-treatment acts through specific enzymes on the lignin polymers. Laccases catalyses the oxidation of the phenolic hydroxyl groups in the lignin substrate, and unspecific alkaline and acid hydrolysis involves lignin and other polysaccharide fractions [19]. The different actions result firstly in evident properties of the liquid recovered after the pre-treatment: light-coloured liquid after biological pre-treatment and dark-coloured liquid after chemical pre-treatment [20,21]. This result is consistent with the decrease of xylan and arabinan fractions due to the unspecific hydrolysis of hemicellulose in alkaline and acid solvents. Accordingly, the specific enzymatic pre-treatment resulted in a high biomass recovery, low lignin removal, and thus a small variation in polysaccharides fractions. It is worth noting that the biological pre-treatment provides large biomass recovery and a small variation of xylans and arabinans, about 40% pentose sugars loss against almost complete (97–99%) loss after acid and alkaline pre-treatments. This result encourages further optimization of the laccase use in the delignification of agro-food wastes whenever pentose sugars can be used as substrates for fermentation processes.

3.2. Effect of Mixing Rate on Kinetics

The rate of glucose production can be strongly affected by the mixing rate due to the heterogeneous nature of the enzymatic hydrolysis of real lignocellulosic biomasses. For this reason, the effect of the mixing rate applied in the stirred reactor has been investigated to assess the minimum mixing rate value above which a constant glucose production rate was observed.

Figure 1 reports the initial hydrolysis reaction rates assessed after 1 h of hydrolysis. Operating conditions were 10% w/v biomass loading, 5 FPU $g_{cellulose}^{-1}$ cellulase activity, and a temperature of 50 °C. The stirring speed was set between 0 and 500 rpm to assess the optimal mixing speed. The optimization was necessary to enhance the liquid–solid mass transfer rate and to investigate the effect on glucose production rate of any other phenomena affected by mixing speed (e.g., biomass aggregates break-up).

As expected, and according to the results reported by Prato et al. [4], the initial hydrolysis rate increased with mixing rate in a range of low mixing rates for all the pre-treated AP samples. This result confirms the strong dependence of apparent glucose production rate on the mixing rate applied to the biomass slurry and thus suggests the necessity to select proper operating conditions (minimum mixing rate) for the quantitative assessment of glucose production kinetics. Tests carried out with NaOH pointed out that the hydrolysis rate for NaOH pre-treated biomass increased with the mixing rate and approached a constant value at mixing rates larger than 180 rpm. The same behaviour was observed in the cases of AP pre-treated with HCl and laccases. The critical value of the mixing rate above which the glucose production rate approached a constant value was 200 and 300 rpm for acid and enzymatic pre-treated AP, respectively. Thus, the minimum values for mixing rates not affecting the apparent glucose production rate during enzymatic hydrolysis of AP samples were 180, 200, and 300 rpm for AP samples pre-treated with alkaline, acid, and biological process, respectively. These values were set for the hydrolysis tests aimed at the assessment of the effects of substrate concentration and product inhibition on the kinetics. Results reported in the next section refer to tests carried out setting the mixing rate at the minimum value to neglect mixing-related phenomena.

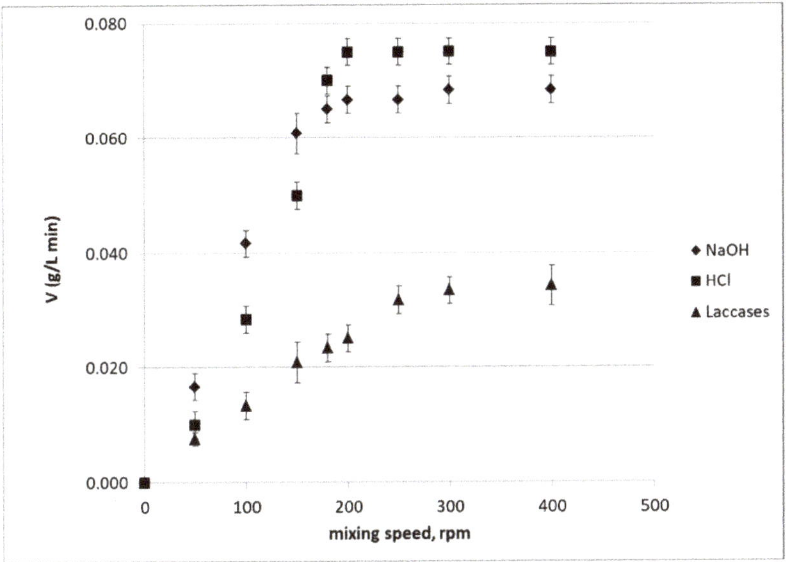

Figure 1. Initial glucose production rates V for pre-treated AP vs. the mixing speed. Enzymatic hydrolysis conditions: solid loading 10% w/v, Cellic®CTec2 concentration 1.2 g L^{-1}, 5 FPU g$_{\text{cellulose}}^{-1}$, 50 °C, pH 4.8, and a hydrolysis time of 1 h.

3.3. Effect of Substrate Concentration on Kinetics

The effect of substrate (glucans) concentration on the glucose production rate has been assessed according to the procedure reported in the Section 2.5 through brief term (1 h) hydrolysis tests on AP samples pre-treated with one of the three abovementioned delignification processes. Figure 2 reports the glucose production rate as a function of substrate (glucans) concentration for all the AP samples. The glucose production rate increased progressively with S and approached a maximum constant value as S was larger than about 20 g L^{-1}. The maximum constant values of the glucose production rates were close to 0.09 g$_{\text{glucose}}$ L$_{\text{solution}}^{-1}$ min^{-1} for AP samples pre-treated with 2% NaOH, 0.070 g$_{\text{glucose}}$ L$_{\text{solution}}^{-1}$ min^{-1} for AP pre-treated with 2% HCl, and 0.04 g$_{\text{glucose}}$ L$_{\text{solution}}^{-1}$ min^{-1} for AP pre-treated with laccases.

Table 2. Kinetic parameters of the pseudo-homogeneous Michaelis and Menten model (Equation (1)) for AP hydrolysis catalysed by Cellic®CTec2.

Pre-Treatment	Glucans (%)	Michaelis and Menten Model	
		K_M (g L^{-1})	V_{Max} (g L^{-1} min^{-1})
NaOH	28.0	5.5 ± 0.6	0.108 ± 0.003
HCl	25.0	4.8 ± 0.6	0.084 ± 0.003
Laccase	22.5	9.0 ± 2.1	0.067 ± 0.006

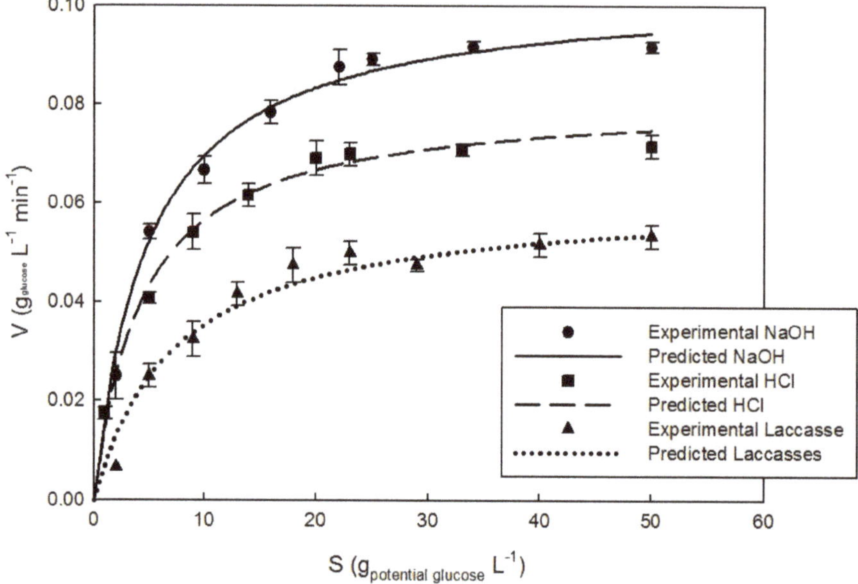

Figure 2. Initial glucose production rate V vs. substrate concentration S for AP pre-treated with 2% NaOH, 2% HCl, and 166 U/mL laccase. Enzymatic hydrolysis conditions: Cellic®CTec2 activity 290 FPU $L_{solution}^{-1}$, 50 °C, pH 4.8, and a hydrolysis time of 1 h. Optimized mixing rate was fixed for each AP sample. Lines are the plot of Equation (1) with parameters reported in Table 2.

According to the procedure proposed by Pratto et al. [4] the substrate concentration has been assumed to be constant during the short tests (1 h). This assumption has been verified after the experimental tests; at small substrate concentrations (S < 10 g L^{-1}) the substrate conversion after 1 h enzymatic hydrolysis was higher than 20% of the potential glucose (see Table 1). This result suggests further future assessments of the initial glucose production rate at low biomass loadings (<1 wt %) through hydrolysis tests shorter than 1 h. Indeed, enzyme adsorption on the substrate and the change in reactivity of the substrate itself [7,12] can influence the reaction rate whenever the large conversion occurs over a short timescale. These phenomena can be monitored under the adopted procedure by proper analysis other than the measurement of glucose concentration in the liquid and by preliminary investigation above the aim of this work (e.g., characterization of enzyme adsorption on biomass surface). Kinetic parameters resulting from regression of data in Figure 2 through Equation (1) are reported in Table 2.

3.4. Effect of Product Inhibition on Kinetics

The effect of glucose inhibition on the kinetics of enzymatic hydrolysis of AP was assessed through long-term tests (72 h) according to the procedure described in Section 2.6. Figure 3 reports experimental data measured during enzymatic hydrolysis of AP pre-treated with NaOH, HCl, and laccases at different biomass loading. The long-term hydrolysis tests were carried out at a 5, 7.5, and 10% w/v biomass loading. According to the reported data, the maximum produced glucose concentration (about 20 g L^{-1}) in the liquid was reached after 48 h when a 10% w/v biomass loading was used. Data regression according to Michaelis and Menten model and Chrastil's model provided the kinetic parameters—V_{max}, K_M, K_i, k, and n—reported in Table 3. The agreement between the experimental data and the proposed models was satisfactory for all the pre-treated biomasses. With regard to the Michaelis and Menten model with product inhibition results showed how the biomass composition resulting from the delignification pre-treatment affected the maximum glucose production rate V_{max}, K_M, and K_i. The highest V_{max} value was reported for the biomass pre-treated with 2% NaOH. This result is in agreement with the low residual lignin content observed in the AP after alkaline pre-treatment (see Table 1) and thus with the expected large fraction of cellulose available for the enzyme productive adsorption [5,10] in this kind of pre-treated AP sample. The highest K_M value was measured after laccases pre-treatment, again as a likely effect of the large lignin content of AP after biological delignification. The type of pre-treatment, and thus the biomass lignin content, did not significantly affect the values of the inhibition parameter K_i. This result is reasonably related to the presence of a mixture of cellulase and β-glucosidase in the enzyme cocktail; indeed, the latter enzymes acts as a homogeneous biocatalyst and thus is not affected by the composition of the biomass. Regarding Chrastil's model, similar observation rise form the comparison of assessed kinetic parameters and the composition of AP samples. The highest n value was reported for hydrolysis of biomass pre-treated with laccases and 2% HCl. An evident trend of the effect of lignin content on n parameter cannot be inferred from the data in Table 3. In any case, according to Chrastil's model (see Section 2.7), the diffusion related resistance caused by the structure of the lignocellulosic substrate seems to me more pronounced for alkaline pre-treated AP samples. As expected, the value of k slightly increased as the AIL of the pre-treated biomass decreased, in accord with results reported by Pratto et al. [13] regarding another biomass typology. The comparison of statistic data suggested that Chrastil's model provided the best fitting and it can be considered for further investigation aimed at bioreactor design. The present study confirms the dependence of the overall sugar production rate on the stirring rate and suggests the adoption of quantitative description of the hydrolysis dynamics under relevant operating conditions.

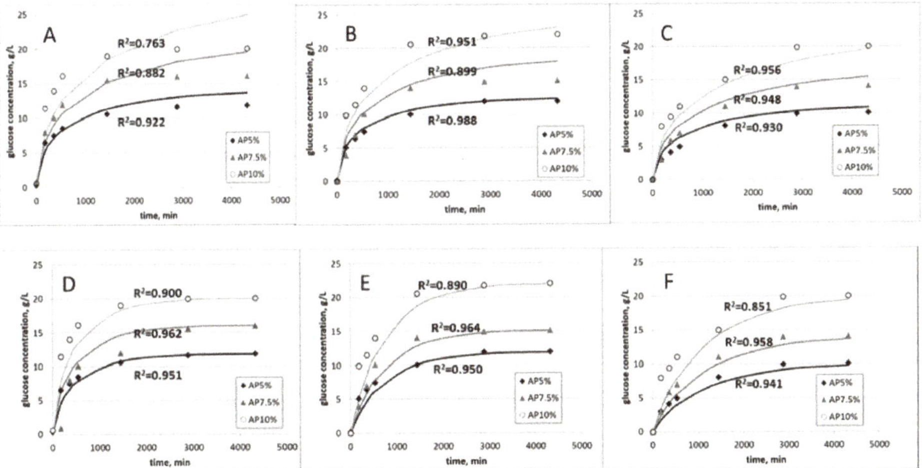

Figure 3. Time resolved concentration of glucose in the liquid phase during enzymatic hydrolysis of the AP at different biomass loading after pre-treatment in 2% NaOH (**A,D**), 2% HCl (**B,E**), and 166 U mL^{-1} laccases (**C,F**). Operating conditions: enzyme concentration set at 828 FPU L^{-1}; optimized mixing rate. Lines mark regression of the experimental data with the Michaelis and Menten with product inhibition model (**A,B,C**) and Chrastil's model (**D,E,F**).

Table 3. Kinetic parameters for Michaelis and Menten with product inhibition and Chrastil's model for enzymatic hydrolysis of AP pre-treated with alkaline (NaOH), acid (HCl), and enzymatic (laccase) delignification processes.

Biomass Pre-Treatment	Glucans (%)	Lignin (%)	M&M with Product Inhibition			Chrastil's Model	
			V_{max} (g L^{-1} min^{-1})	K_M (g L^{-1})	K_i (g L^{-1})	k (L g^{-1} min^{-1})	n (-)
NaOH	28.0	5.0	0.30	5.5	0.16 ± 0.02	$2 \cdot 10^{-3} \pm 1.2 \cdot 10^{-4}$	0.41 ± 0.02
HCl	25.5	9.0	0.24	4.8	0.20 ± 0.03	$1.8 \cdot 10^{-3} \pm 1.2 \cdot 10^{-4}$	0.63 ± 0.03
Laccase	22.5	14.5	0.19	9.0	0.31 ± 0.03	$1.1 \cdot 10^{-3} \pm 8 \cdot 10^{-5}$	0.58 ± 0.03

The assessed kinetic parameters were compared with the results reported in the literature (Table 4). To the best of our knowledge, the data comparison was limited to few studies of the overall hydrolysis rate of raw lignocellulosic biomass. Carvalho et al. [14] studied the enzymatic hydrolysis rate of sugarcane bagasse after a sequence of steam explosion and alkaline pre-treatment. Pratto et al. [4] investigated the enzymatic hydrolysis rate of sugarcane straw after a sequence of hydrothermal and alkaline (4% NaOH) pre-treatment. Data reported in the literature were compared with those assessed in the present paper related to alkaline pre-treatment of AP (Table 4). The n values reported for steam-exploded sugarcane bagasse is larger than those reported for the other two biomasses; moreover, the same n value has been assessed for two different substrates (sugarcane straw and AP) subjected to the same delignification pre-treatment (alkaline hydrolysis). These values of the n parameters seem to bc in reasonable agreement with the structures of steam exploded and chemically pre-treated biomass. Indeed, structural modification after steam explosion can positively affect the limitation by diffusion of the heterogeneous enzymatic process [4,5]. The k values assessed in the present paper are remarkably larger than those reported by Carvalho et al. [14] and by Pratto et al. [4], and this result is related with the strong dependence of the maximum rate of hydrolysis on the biomass composition even if it is not related to the lignin content (similar residual lignin in Table 4 for all biomass types). Thus, the variation of k parameter can be due to the different biomass investigated and to the overall structural and composition variation after delignification pre-treatment of AP [18].

Table 4. Kinetic parameters for Chrastil's model reported in the literature for enzymatic hydrolysis of real lignocellulosic biomass.

Biomass	Pre-Treatments	Residual Lignin (%)	Chrastil's Model		References
			k (L g^{-1} min^{-1})	n (-)	
Sugarcane Bagasse	121 °C 30 min in 4% NaOH + Steam explosion	5.4	$2 \cdot 10^{-5}$	0.6	[12]
Sugarcane Straw	195 °C 10min + 121 °C 30 min in 4% NaOH	5.7	Between $3.52 \cdot 10^{-5}$ and $7.4 \cdot 10^{-5}$	0.4	[13]
Apple Pomace	121 °C 30 min in 2% NaOH	5.0	$2 \cdot 10^{-3}$	0.4	Present work

4. Conclusions

A procedure for the assessment of the overall kinetics of glucose production from apple pomace enzymatic hydrolysis catalysed by cellulase cocktails was applied using a commercial enzyme cocktail (Cellic®CTec2). The effect of mixing rate on the observed overall glucose production rate was confirmed for AP hydrolysis according to the results reported in the literature for other biomasses [4]. According to the reported results, faster kinetics were observed during hydrolysis of biomass characterized by a low lignin content. Since the residual lignin content was a result of the delignification pre-treatment (minimized after the alkaline pre-treatment), the effect of pre-treatment on the enzymatic hydrolysis rate (glucose production) was quantified by the assessed kinetic parameters. The analysis of the glucose production dynamics pointed out that both the adopted kinetic models provided a satisfactory description of the process. Chrastil's model was the most effective and can be suggested for the rational design of enzymatic hydrolysis reactors supplemented with AP.

The present study is the starting point of a wider analysis aimed at the generation of quantitative tools based on the overall rate of enzymatic hydrolysis of real biomass. These data enable the design and the optimization of enzymatic hydrolysis reactors and their operating conditions depending on the different possible biomass compositions (sugars and lignin content) resulting after the pre-treatment step. Because the obtained results are in agreement with the results by Pratto et al. [13] and because of the different biomasses used by Pratto et al. [13] and in the present study (see Table 4), the adopted procedure can be successfully extended to several potential feedstock for lignocellulose saccharification (mainly agro-food wastes for second generation bio-based fuels and chemicals production).

Supplementary Materials: The following are available online at http://www.mdpi.com/1996-1073/13/5/1051/s1.

Author Contributions: Conceptualization, M.E.R. and A.P.; Methodology, M.E.R. and A.P.; Software, I.D.S.; Validation, M.E.R. and A.P.; Formal Analysis, M.E.R.; Investigation, A.P.; Resources, A.M. and M.E.R.; Data Curation, A.P.; Writing-Original Draft Preparation, M.E.R. and A.P.; Writing-Review & Editing, M.E.R. and A.M.; Visualization, A.P.; Supervision, M.E.R. and A.M.; Project Administration, A.M. and M.E.R; Funding Acquisition, A.M. and M.E.R. All authors have read and agreed to the published version of the manuscript.

Funding: This research was funded by European Union's Horizon 2020 research and innovation program by the project "Waste2Fuels Sustainable production of next generation biofuels from waste streams" grant number N°654623. Check carefully that the details given are accurate and use the standard spelling of funding agency names at https://search.crossref.org/funding, any errors may affect your future funding.

Acknowledgments: The authors thank Alessio Occhicone for his participation to parts of the experimental tests and data analysis.

Conflicts of Interest: The authors declare no conflict of interest. The funders had no role in the design of the study; in the collection, analyses, or interpretation of data; in the writing of the manuscript, and in the decision to publish the results.

References

1. Parajuli, R.; Dalgaard, T.; Jørgensen, U.; Adamsen, A.P.; Trydeman Knudsen, M.; Birkved, M.; Gylling, M.; KofodSchjørring, J. Biorefining in the prevailing energy and materials crisis: A review of sustainable pathways for biorefinery value chains and sustainability assessment methodologies. *Renew. Sustain. Energy Rev.* **2015**, *43*, 244–263. [CrossRef]

2. Arora, A.; Banerjee, J.; Vijayaraghavan , R.; Mac Farlane, D.; Patti, A.F. Process design and techno-economic analysis of an integrated mango processing waste biorefinery. *Ind. Crops Prod.* **2018**, *116*, 24–34. [CrossRef]

3. Zhang, L.; Freitas dos Santos, A.C.; Ximenes, E.; Ladisch, M. Proteins at heterogeneous (lignocellulose) interfaces. *Curr. Opin. Chem. Eng.* **2017**, *18*, 45–54. [CrossRef]

4. Pratto, B.; Alencar de Souza, R.B.; Sousa, R., Jr.; Gonçalves da Cruz, A.J. Enzymatic Hydrolysis of Pretreated Sugarcane Straw: Kinetic Study and Semi-Mechanistic Modeling. *Appl. Biochem. Biotechnol.* **2016**, *178*, 1430–1444. [CrossRef] [PubMed]

5. Mohammad, B.T.; Al-Shannag, M.; Alnaief, M.; Singh, L.; Singsaas, E.; Alkasrawi, M. Production of multiple biofuels from Whole Camelina Material: A renewable energy crop. *BioResources* **2019**, *13*, 4870–4883.

6. Tan, U.L.; Yu, E.K.C.; Campbell, N.; Saddler, J.N. Column cellulose hydrolysis reactor: An efficient cellulose hydrolysis reactor with continuous cellulase recycling. *Appl. Microbiol. Biotechnol.* **1986**, *25*, 250–255. [CrossRef]

7. Gusakov, A.; Sinitsyn, A.P.; Klyosov, A.A. A Theoretical Comparison of the Reactors for the Enzymatic Hydrolysis of Cellulose. *Biotechnol. Bioeng.* **1987**, *29*, 898–900. [CrossRef] [PubMed]

8. Xue, Y.; Jameel, H.; Phillips, R.; Chang, H. Split addition of enzymes in enzymatic hydrolysis at high solids concentration to increase sugar concentration for bioethanol production. *J. Ind. Eng. Chem.* **2012**, *18*, 707–714. [CrossRef]

9. González Quiroga, A.; Bula Silvera, A.; Vasquez Padilla, R.; da Costa, A.C.; Maciel Filho, A.C. Continuous and semicontinuous reaction systems for high-solids enzymatic hydrolysis of lignocellulosic. *Braz. J. Chem. Eng.* **2015**, *32*, 805–819. [CrossRef]

10. Zhang, X.; Yuan, Q.; Cheng, G. Deconstruction of corncob by steam explosion pretreatment: Correlations between sugar conversion and recalcitrant structures. *Carbohydr. Polym.* **2017**, *156*, 351–356. [CrossRef] [PubMed]

11. Jeoh, T.; Cardona, J.M.; Karuna, N.; Mudinoor, A.R.; Nill, J. Mechanistic Kinetic Models of Enzymatic Cellulose Hydrolysis—A Review. *Biotechnol. Bioeng.* **2017**, *114*, 7. [CrossRef] [PubMed]

12. Bansal, P.; Hall, M.; Realff, M.J.; Lee, J.H.; Bommarius, A.S. Modelling cellulase kinetics on lignocellulosic substrates. *Biotechnol. Adv.* **2009**, *27*, 833–848. [CrossRef] [PubMed]

13. Carrillo, F.; Lis, M.J.; Colom, X.; López-Mesas, M.; Valldeperas, J. Effect of alkali pretreatment on cellulose hydrolysis of wheat straw: Kinetic study. *Proc. Biochem.* **2005**, *40*, 3360–3364. [CrossRef]

14. Carvalho, M.L.; Sousa, R., Jr.; Rodríguez-Zúñiga, U.F.; Suarez, C.A.G.; Rodrigues, D.S.; Giordano, R.C.; Giordano, R.L.C. Kinetic study of the enzymatic hydrolysis of sugarcane bagasse. *Braz. J. Chem. Eng.* **2005**, *30*, 437–447. [CrossRef]

15. Procentese, A.; Raganati, F.; Olivieri, G.; Russo, M.E.; Rehmann LMarzocchella, A. Deep eutectic solvents pretreatment of agro industrial food waste. *Biotech. Biofuels* **2018**, *11*, 37. [CrossRef] [PubMed]

16. Adney, B.; Baker, J. *Measurement of Cellulase Activities*; Report No.: NREL/TP-510-42628; National Renewable Energy Laboratory: Denver, CO, USA, 2008.

17. Sluiter, A.; Hames, B.; Ruiz, R.; Scarlata, C.; Sluiter, J.; Templeton, D.; Crocker, D. *Determination of Structural Carbohydrates and Lignin in Biomass*; Report No.: NREL/TP-510-42618; National Renewable Energy Laboratory: Denver, CO, USA, July 2011.

18. Chrastil, J. Enzymic product formation curves with the normal or diffusion limited reaction mechanism and in the presence of substrate receptors. *Int. J. Biochem.* **1988**, *20*, 683–693. [CrossRef]

19. Kumar, P.; Barrett, D.M.; Delwiche, M.J.; Stroeve, P. Methods for pretreatment of lignocellulosic biomass for efficient hydrolysis and biofuel production. *Ind. Eng. Chem. Res.* **2009**, *48*, 3713–3729. [CrossRef]

20. Niglio, S.; Procentese, A.; Russo, M.E.; Piscitelli, A.; Marzocchella, A. Integrated enzymatic pretreatment and hydrolysis of apple pomace in a bubble column bioreactor. *Biochem. Eng. J.* **2019**, *150*, 107306. [CrossRef]

21. Procentese, A.; Raganati, F.; Olivieri, G.; Russo, M.E.; Marzocchella, A. Pre-treatment and enzymatic hydrolysis of lettuce residues as feedstock for bio-butanol production. *Biomass. Bioenergy* **2017**, *96*, 172–179. [CrossRef]

Article

Extraction Behaviors of Lignin and Hemicellulose-Derived Sugars During Organosolv Fractionation of Agricultural Residues Using a Bench-Scale Ball Milling Reactor

Tae Hoon Kim [1,2], Hyun Kwak [1], Tae Hyun Kim [2,*] and Kyeong Keun Oh [1,3,*]

[1] R&D Center, SugarEn Co., Ltd., Yongin 16890, Gyeonggi-do, Korea; thkim@sugaren.co.kr (T.H.K.); hkwak@sugaren.co.kr (H.K.)
[2] Department of Materials Science and Chemical Engineering, Hanyang University, Ansan 15588, Gyeonggi-do, Korea
[3] Department of Chemical Engineering, Dankook University, Youngin 16890, Gyeonggi-do, Korea
* Correspondence: hitaehyun@hanyang.ac.kr (T.H.K.); kkoh@dankook.ac.kr (K.K.O.); Tel.: +82-31-400-5222 (T.H.K.); +82-31-8005-3548 (K.K.O.)

Received: 15 November 2019; Accepted: 8 January 2020; Published: 10 January 2020

Abstract: Ethanol organosolv fractionation combined with ball milling was conducted on three major agricultural residues: Rice husk (RH), rice straw (RS), and barley straw (BS). The highest lignin extraction yields of RH, RS, and BS were 55.2%, 53.1%, and 59.4% and the purity of lignin recovered was 99.5% for RH and RS, and 96.8% for BS, with similar chemical characteristics, i.e., low molecular weight distributions (1453–1817 g/mol) and poly dispersity index (1.15–1.28). However, considering the simultaneous production of hemicellulose-derived sugars, distinctive fractionation behaviors were shown for the three agricultural residues. The highest hemicellulose-derived sugar yield was 73.8% when RH was fractionated at 170 °C for 30 min. Meanwhile, very low sugar yields of 31.9% and 35.7% were obtained from RS and BS, respectively. The highest glucan-to-glucose conversion yield from enzymatic hydrolysis of fractionated RH reached 85.2%. Meanwhile, the enzymatic digestibility of the fractionated RS and BS was 60.0% and 70.5%, respectively. Consequently, the fractionation efficiency for RH can be improved with fine refinement. For the case of RS, other fractionation process should be applied to achieve effective fractionation performance.

Keywords: biomass; pretreatment; biofuel; biorefinery; sugar-decomposed

---

1. Introduction

Due to the enormous challenges posed by increasingly severe climate change, lignocellulosic biomass, a renewable and sustainable resource, has been considered as a promising alternative to the finite oil reserves as it can be used to produce alternative fuels and chemicals [1]. The serious environmental problems associated with the rapidly increasing use of fossil fuels are currently increasing the need for greater use of renewable resources, such as lignocellulosic biomass. Because lignocellulosic biomass is abundant and reproducible, various efforts to use it for the production of industrial bio-based products have attracted great attention over the past decades [2,3].

Recently, numerous efforts have been made to study how to effectively utilize various lignocellulosic biomass, including those obtained from agricultural residues, forest residues, and energy crops [4]. It is assumed that one of the aforementioned various agricultural resources must offer an important renewable source that is a byproduct of crop cultivation and has a structure that is more accessible by enzymes or other chemicals compared to other biomass and is therefore expected to play an important role in future biorefineries. At present, large amounts of agricultural products are cultivated all over the world, and appropriate planning is required to manage the large amounts of agricultural residues produced. Some of the agricultural byproducts produced are used in animal feeding, animal bedding, soil mulching, composting, and household fuels, but most of it is generally incinerated and disposed on-farm, which has a negative impact on human health and the environment [5].

According to a Statista (Hamburg, Germany) report on worldwide 2017/2018 paddy rice production, there are two major regions of South and East Asia. In detail, China and India are two major rice producers in the world. In 2018, China's rice production was more than 290 million metric tons (MMT), accounting for one-third of the world's rice production (495.9 MMT worldwide). For reference, Chinese rice production is nearly twice that of India (169.5 MMT) and hundreds of times that of Korea (5.3 MMT) [6]. The global production volume of barley amounted to 142.4 MMT in the 2017/2018 crop year. This statistic shows the European Union (EU), the leading producer, produced over 56 MMT of barley that year [7]. A factsheet from FAO (Food and Agriculture Organization of the United Nations) on rice production reports a "rice straw (lignocellulose)-to-grain (starch)" ratio of 1.1 for most currently planted rice varieties, and rice husk accounts for about 20 wt.% of grain, and is one of the major agro-industrial by-products produced worldwide [2]. Although the straw-to-grain ratio of barley is affected by factors, such as the plant height, spike length, and grain weight, it is generally considered to be approximately 0.75 [8].

The agricultural residues mentioned above, rice straw (RS), rice husk (RH), and barley straw (BS), are fibrous residues generated after the harvest of the principal grains used by mankind, and they are produced at a greater amount than other biomass types. It is a raw material that does not compete with food and is relatively easy to collect and transport using existing infrastructure.

One of the more efficient uses of agricultural by-products (lignocellulosic biomass) into fuels and chemicals is through the biorefining process. Agricultural by-product feedstock is separated into three main constituents (cellulose, hemicellulose, and lignin), each of which is converted into fuel, chemicals, and other materials through a variety of thermochemical and biochemical processes [2]. However, the physicochemical linkage of cellulose–hemicellulose–lignin is a major factor affecting the rate and yield of enzymatic saccharification and makes the biodegradation of lignocellulosic very slow or extremely difficult in nature. It is believed that the extensive and complex interactions that occur in the molecular and macromolecular networks between cellulose, hemicellulose, and lignin in the plant cell wall matrix contribute to the difference in their reactivity during pretreatment. Information on the composition of the plant cell walls of various plant resources is very importance in developing process technologies that make the best use of lignocellulosic biomass and the added value through mechanical, chemical, and biological pretreatment. [1].

Typically, 60% to 75% (based on weight) of lignocellulosic biomass is composed of polysaccharides (cellulose and hemicellulose), which can be hydrolyzed to produce five different monosaccharides, such as glucose, xylose, arabinose, galactose, and mannose, and then further utilized as a substrate in microbial fermentative for the production of value-added products [9]. In the conversion of biomass, hydrolysis of polysaccharides to monosaccharides, in particular cellulose to glucose is a major hurdle to achieve complete utilization of biomass. Following cellulose, hemicellulose is a macromolecule composed mainly of xylose, but other sugars with a lower molecular weight than that of cellulose, such as mannose, galactose, rhamnose, and arabinose, are also contained. These water–carbohydrates can be converted into various industrial chemicals and fuels, such as ethanol, hydrocarbon substitutes, and starting monomers, for the production of biopolymers [10,11]. In addition to cellulose and hemicellulose, lignin is a complex macromolecule that accounts for a significant portion of the plant cell wall, and lignin is the second most abundant biopolymer on earth after cellulose [12]. Basically, lignin monomers are divided into three aromatic alcohols: *p*-coumaryl, coniferyl, and sinapyl [13]; these three-unit monomers are linked by various types of ethers and carbon–carbon bonds to form large polymers [14]. Despite the many benefits of lignin, such as abundance and high potential, i.e., high carbon content and biodegradability, purification of lignin from lignocellulosic biomass still remains difficult and current process technology development is still unsatisfactory. Thus, the efficient separation and utilization of lignin remains a challenging part of research and investment in bioremediation processes despite its high potential as a carbon source for future biochemical industries [15,16].

Therefore, the development of an effective process that separates the lignin and cellulose fractions from lignocellulosic biomass is necessary for an economically feasible biomass-refining industry. The main requirement for an economically viable process is to generate highly pure lignin, and minimal waste compounds using efficient processing [17]. In general, the fractionation methods for lignocellulosic biomass mainly include physical, chemical, and biological methods and/or their combinations [18]. These varied processes have different impacts on the structure of the lignocellulosic biomass after processing, and result in a significant impact on the downstream biomass conversion process in terms of sugar recovery yields, toxicity of hydrolysates, enzymatic hydrolysis rate and yield, microbial fermentation yield, and waste treatment costs incurred [19]. Among them, thermochemical fractionation results in the production of severe sugar-degraded products and various toxic compounds in the liquid hydrolysate, such as 5-hydroxymethylfurfural (HMF), furfural, formic acid, and acetic acid [20]. Organosolv fractionation has been known to provide a high recovery yield of lignin-derived and hemicellulose-derived sugars; however, fine optimization of the reaction conditions used for the hydrolysis is important to ensure high monomeric sugar recovery and minimal release of potential inhibitors to microbial fermentation. Therefore, an effective lignin separation process needs to be developed to reduce the toxicity of hemicellulosic hydrolysate during fractionation, which makes the high value-added material production from biorefining more economically viable.

In recent years, the conventional ball milling pretreatment method has received a lot of interest as an effective physical pretreatment method, combining a proven chemical pretreatment and a ball milling pretreatment process, and has been proposed to increase the enzymatic hydrolysis of biomass. [18,21–24]. Previous studies have reported that ball milling pretreatment methods increase the accessible surface area of biomass and effectively reduce the crystallinity and polymerization degree of cellulose [25–27]. Among the chemical pretreatment methods, organosolv fractionation using organic solvents is known as an efficient method for effectively depolymerizing lignin from lignocellulosic biomass. In general, the organosolv fractionation method employs organic solvent, mainly ethanol, which can be recovered and reused [28,29]. The advantage of using organosolv fractionation method is that highly pure lignin can be obtained in a liquid form, which can be used for high-value products, such as adhesives, fibers, and biodegradable polymers [30,31]. In our previous work [32], ball milling combined with organosolv fractionation was reported as an effective method in terms of increased glucan retention in solid residues and lignin recovery, which resulted in significantly improved enzyme digestibility on herbaceous biomass. After the organosolv pretreatment, a cellulose-rich solid cake was obtained, whereas most of the lignin and part of the hemicellulose were recovered by dissolution in an organic solvent and recovered. Obviously, lignin removal leads to an increased surface area, making cellulose more accessible to enzymes [33].

In this study, ethanol organosolv fractionation combined with ball milling was conducted on various agricultural residues. Three major agricultural residues, RH, RS, and BS, were chosen to characterize the fractionation behaviors. The effectiveness of the fractionation process was evaluated in terms of the glucan retention, hemicellulose-derived sugar recovery, and lignin recovery yield. The enzymatic digestibility of the residual solids and byproduct formation were also determined.

2. Materials and Methods

2.1. Raw Materials

Three different agricultural residues—RH, RS, and BS—were provided by the National Institute of Crop Science (NICS), Rural Development Administration (RDA, Wanju, Jeolla-bukdo, Korea). The samples were ground by a commercial blender (Blender 7012s, Waring Commercial, CT, USA) and then sieved to a nominal size of 14–45 mesh (from 0.36 to 1.40 mm). The ground biomass was placed in a convection oven at 45 ± 5 °C for 48 h, and then stored in an automatic dehumidification desiccator until use. The average moisture content of the dried residues was 4.0% to 5.5% during the experiments. Ethanol (cat. no. E7023), sulfuric acid (cat. no. 258105), tetrahydrofuran (cat. no. 401757), pyridine (cat. no. 270970), chloroform-d (cat. no. 151858), cyclohexanol (cat. no. 105899), chromium (III) acetylacetonate (cat. no. 574082), and 2-chloro-4,4,5,5-tetramethyl-1,2,3-dioxaphospholane (cat. no. 447536) were purchased from Sigma–Aldrich Korea (Yongin, Gyeongdo, Korea).

2.2. Experimental Setup and Operation

An illustration of a bench-scale ball milling reactor is shown in Figure 1. The ball milling reactor of this system was constructed using SS-316L material with a wall thickness of 0.01 m (inner diameter 0.30 m, depth 0.42 m, and internal volume 30 L).

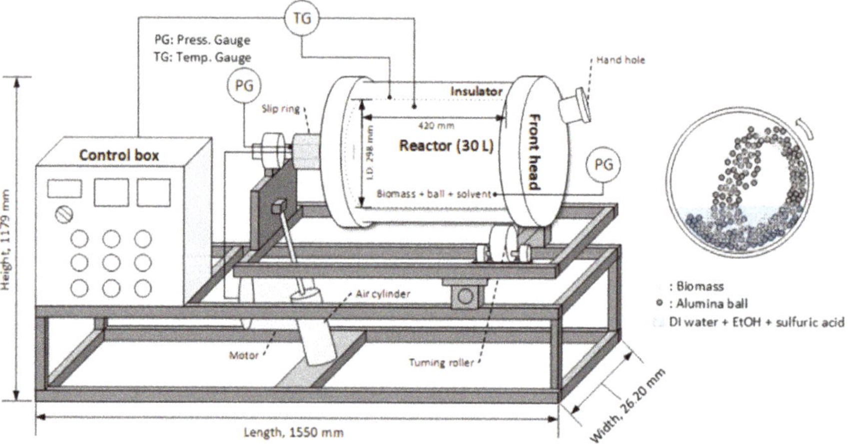

Figure 1. Diagram of the temperature-controlled and tilted bench-scale (30 L) ball milling reactor system.

The reactor system can be operated at pressures up to 1.96 MPa, 200 °C temperature, and 6.28 rad/s, and metal spiral gaskets were installed at the connection between the reactor and the head at the front and rear to maintain the pressure. The reactor can be rotated using a 3-phase, 380 V, 0.37 kW-riven motor, with a slip ring with a rotating connector that can supply power or signals to the rotating equipment without twisting the power cables. The front head was equipped with a 0.075 m diameter hole for supplying reactants, biomass, balls, and solvents and discharging the product, and the reactor was heated and controlled by a 10 kW mantle-type electric heater. The high-temperature high-pressure ball mill reactor has a tilting system, and because of its heavy weight, the reactor body is operated by placing the reactor on a frame with a rotating roller. The spherical shape alumina ball (HD, 0.01 m diameter, and 3600 kg/m^3 density) was purchased from NIKKATO Co. (Osaka, Japan). Before operating the reactor, alumina balls, biomass, and ethanol (60% *v/v*, 0.25% sulfuric acid *w/v*) were put together in the reactor, where the ratio of the ball:biomass:solvent was 30:1:10 (*w/w/v*). The reactor was preheated while rotating for 1 h until the target temperature (170 or 190 °C) was reached. When the target temperature was reached, the reaction was maintained while maintaining the temperature for a reaction time of 30, 60, or 90 min. Upon completion of the reaction, the sugar and lignin in the liquid sample were analyzed for the sugar yield and lignin separation. For the lignin content analysis, the sample was evaporated at 55 °C for 3 h, and then diluted three times with DI (deionized) water to precipitate the lignin. Water was then evaporated at 55 °C for 4 h, and acid hydrolysis using sulfuric acid was performed by adding highly concentrated sulfuric acid (72%), finally making 4.0% sulfuric acid to analyze the sugar and lignin in the sample, which was then subjected to an autoclave for 1 h at 121 °C. The lignin precipitate obtained above was washed with ethanol (60%, *v/v*) and water, dried at

45 °C in a convection oven, and subjected to compositional analysis. All experiments and analyses were repeated three times.

2.3. Characterization of Organosolv-Separated Lignin

Characterization of organosolv-separated lignin was performed and the average molar mass (M_n), weight average molar mass (M_w), and poly-dispersity (PD) of the organosolv-separated lignin were determined by GPC (gel permeation chromatography, Ultimate 3000 model, Thermo Fisher Scientific Inc., Waltham, USA). For molecular weight measurement, 3.0 mg of the acetylated lignin sample were dissolved in 2.0 mL of tetrahydrofuran (THF) and then filtered using a 0.45-μm PTFE (polytetrafluoroethylene) syringe filter to remove impurities. A Shodex column (KF-806L model) with a Shodex guard column (KF-G model) and refractive index (RI) detector (RefractoMax 520 model) using THF as a mobile phase (1.0 mL/min) with injection volumes of 20 μL were used in the GPC system. For the FT-IR (Fourier transform infrared) spectroscopy analysis using the IRSpirit-L/T model (Shimadzu Inc., Kyoto, Japan), the attenuation total reflectance (ATR) method was used to determine the characteristic absorption peak of the chemical functional groups in organosolv-fractional lignin. Mid-IR spectra were collected by averaging 40 scans from 4000 to 500 cm^{-1} at a resolution of 1 cm^{-1}. ^{31}P NMR (AVANCE 600, Bruker, Germany) spectra were collected at 242.88 MHz for 256 scans with a 2 s pulse delay. For quantitative ^{31}P NMR analysis, 20 mg of lignin sample were weighed and dissolved into 400 μL of solution A and 150 μL of solution B in a 5-mL vial: Solution A (pyridine:chloroform-d ($CDCl_3$) = 1.6:1 (*v/v*), solution B (mixture of solution A (25 mL), cyclohexanol (100 mg), and chromium (III) acetylacetonate (90 mg)). The prepared liquid sample was mixed well for 5 min using a vortex mixer, and 70 μL of 2-chloro-4,4,5,5-tetramethyl-1,2,3-dioxaphospholane (TMDP) were then added to the solution and placed in the ^{31}P NMR system.

2.4. Solid and Liquid Composition Analysis of Untreated and Pretreated Samples

The carbohydrates and lignin contents in both solid and liquid samples were determined by following the NREL (National Renewable Energy Laboratory, Golden, CO, USA) LAPs (laboratory analytical procedures) [34,35]. The extractive analysis was firstly carried out by a two-step extraction method using water followed by ethanol. For the composition analysis of extractive-free and fractionated solids, a two-step acid hydrolysis following NREL-LAP was applied to determine the carbohydrate and lignin contents. To quantify the monomeric sugars and organic acid in the samples, an HPLC (high-performance liquid chromatography) system (LC-10A, Shimadzu Inc., Kyoto, Japan) with a refractive index detector (RID-10A, Shimadzu Inc., Kyoto, Japan) was used. HPLC equipped with an Aminex HPX-87P carbohydrate column (Bio-Rad Inc., Hercules, USA) was used for sugar analysis and HPLC-grade highly purified water was used as the mobile phase at a flow rate of 0.6 mL/min. For the Aminex HPX-87P column, the operating temperature of the column was maintained at 85 °C. On the other hand, Aminex HPX-87H (cat. No. 125-0098, Bio-Rad Inc., Hercules, USA) was used for organic acid analysis, where 5 mM sulfuric acid was used as a mobile phase at a volumetric flow rate of 0.6 mL/min. The column temperature was kept at 65 °C for organic acid analysis.

2.5. Enzymatic Digestibility Test

The enzymatic digestibility of untreated and treated solids was measured following NREL-LAP [36]. Enzymatic digestibility tests of samples were carried out using a 250-mL Erlenmeyer flask, which was incubated and shaken at 50 °C and 150 rpm in a shaking incubator (model BF-175SI, BioFree Co., Ltd., Seoul, Korea). Commercial cellulase enzyme (Cellic®CTec2, Novozymes Inc., A/S Bagsvaerd, Denmark) was used at the enzyme loading of 15 FPU/g-glucan. To prepare the testing sample in the 250-mL flask with 100 mL of working volume, the initial glucan loading was 1.0% (*w/v*) with 50 mM sodium citrate buffer (pH = 4.8). As antibiotics prevent microbial contamination, 1.0 mL of sodium azide (20 mg sodium azide /mL) was added. Samples were taken at appropriate sampling times (6, 12, 24, 36, 48, 60, and 72 h) and then were subjected to HPLC analysis using an HPX-87H column (Bio-Rad

Laboratories Inc., Hercules, CA, USA). The total released glucose in the flask at 72 h of hydrolysis was used to calculate the enzymatic digestibility based on the glucan content in the untreated sample.

3. Results and Discussion

3.1. Compositions of Raw Agricultural Residues Used in This Study

Compositional analysis was performed on three lignocellulosic biomasses through five individual replicates. The compositional analysis results of the three untreated biomasses are presented in Table 1. For all the samples, the error values are represented as standard deviations. Among the three samples, BS was analyzed to have the highest carbohydrate content: 41.5% of glucan and 21.4% of hemicellulose-derived sugar polymer, i.e., xylan, arabinan, and galactan. On the other hand, RS and RH contained 32.9% and 35.6% of glucan and 20.6% and 15.3% of hemicellulose-derived sugar, respectively. The carbohydrate contents of the three biomass samples were found to be consistent with previous reports [37–39]. Unlike the carbohydrate contents, RH had the highest lignin (acid soluble + acid insoluble) content of 23.1%, compared to the 12.6% and 18.6% observed for RS and BS, respectively. It is also noteworthy that RS and RH showed higher ash contents (14.1% and 15.7%, respectively) than BS (7.4%). This is in line with the current findings on the extraction of minerals (silica) from rice byproducts [40–42]. These chemical compositional differences were assumed to be due to the properties of each residue. Therefore, it is important to understand the fractionation reaction behaviors upon the properties of each biomass types.

Table 1. Chemical compositions of untreated agricultural residues based on an oven-dry biomass.

Components		Composition (wt. %)		
		Rice Straw	**Rice Husk**	**Barley Straw**
Carbohydrates	Glucan	32.9 ± 0.5	35.6 ± 0.8	41.5 ± 0.8
	Xylan	16.2 ± 0.3	13.6 ± 0.4	18.1 ± 0.5
	Galactan	1.5 ± 0.1	-	0.7 ± 0.1
	Arabinan	3.1 ± 0.1	1.7 ± 0.0	2.6 ± 0.1
Lignins	Acid insoluble lignin	11.7 ± 0.0	22.7 ± 0.0	14.9 ± 0.2
	Acid soluble lignin	0.9 ± 0.0	0.7 ± 0.2	3.7 ± 0.0
Extractives	Water	10.9 ± 2.6	6.6 ± 0.1	11.6 ± 0.8
	Ethanol	3.1 ± 0.3	1.2 ± 0.0	1.3 ± 0.1
	Ash	14.1 ± 0.0	15.7 ± 0.2	7.4 ± 0.1
Total		94.4	97.8	101.8

Values are expressed as mean and standard deviation (number of replicates, n = 15).

3.2. Lignin Extraction Yields and Purities for Various Fractionation Conditions

One important component in lignocellulosic biomass is lignin, which is one of the most abundant and renewable aromatic resources in the world [43]. The fractionation reaction conditions in the presence of the appropriate catalyst may affect the lignin's solubilization and extraction behavior, i.e., this fractionation process should be able to break the intermolecular bonding between lignin and hemicellulose and therefore give relatively pure lignin [44].

First, chemical fractionation using organosolv was conducted to extract lignin from three agricultural residues (RH, RS, and BS) by applying various temperatures and reaction times. The fractionation reaction was tested at elevated temperatures (170–190 °C) for extended reaction times (30–90 min), and the solubilization of chemical components during the organosolv fractionation reaction was investigated, as shown in Figure 2. For RH, the lignin recovery yields in the liquid hydrolysate ranged from 43.8% (170 °C, 30 min) to 55.2% (190 °C, 90 min). For the tested reaction conditions, lignin recovery showed a significantly broad distribution, in the ranges of 0.6% to 53.1% for

RS and 16.3% to 59.4% for BS. In all three residues, the lignin extraction yield tended to increase for increased reaction severity (at higher temperatures and for longer reaction times). It was observed that the highest yields of RH, RS, and BS were 55.2%, 53.1%, and 59.4% at 190 °C for 90 min of reaction, respectively.

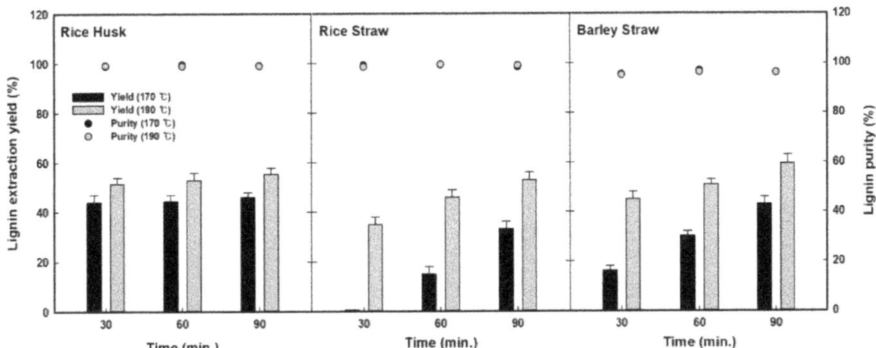

Figure 2. Changes in lignin extraction yield and purity profiles on three agricultural residues (rice husk, rice straw, and barley straws) with various fractionation conditions.

Different lignin extraction behaviors were observed for various reaction conditions. First, in the case of RH, no significant change was observed for lignin extraction for the various reaction conditions. On the other hand, in the case of RS, significant lignin extraction occurred at 170 °C and, as the temperature was increased to 190 °C, the lignin extraction yield increased further to 53.0%, and the effect of the reaction time was also very significant. In particular, in the fractionation of BS, the yield of lignin extraction increased monotonically depending on the severity of the reaction conditions, which was unlikely in the fractionation of RH or RS. On the other hand, highly pure lignin was recovered from all three residues, with a 95.4% to 96.8% content in BS, slightly lower than that of the other residues (98.9% to 99.5% in RH and 98.5% to 99.5% in RS). It was speculated that the precipitation of lignin by the addition of water to the fractionated liquid resulted in this high purity of the lignin because of its low solubility in water.

3.3. Compositional Changes in Solid Residues after Fractionation

The compositional changes in the treated solid residues observed for various tested reaction conditions are presented in Table 2. Under the tested fractionation conditions, the solid remaining after fractionation (weight (g) of the remaining treated solid residue/weight (g) of untreated solid input) were 60.7% to 58.6% at 170 °C and 55.0% to 48.5% at 190 °C for the raw RH as the reaction time increased from 30 to 90 min. The glucan contents in the treated RH solid increased from 35.6% to 54.0% to 55.4% at 170 °C and increased to 55.6% to 54.1% at 190 °C with an increased reaction time, which corresponded to 32.8% to 32.4% and 30.5% to 26.2% (the values were calculated considering the remaining solid), respectively, based on the untreated RH. In the case of RH, the glucan content decreased slightly as the reaction time increased at the reaction temperature of 190 °C, which may indicate that the decomposition of glucan is somewhat sensitive to the reaction time. Meanwhile, the XMG (Xylan + Mannan + Galactan) contents decreased sharply from 13.6% of untreated RH to 4.6% to 3.3% at 170 °C and 1.9% to 0.7% at 190 °C in treated RH as the reaction time increased, which corresponded to 2.8% to 1.9% and 1.0% to 0.4% of the untreated RH, respectively. In addition, the residual lignin contents decreased from 22.7% of untreated RH to 12.3% to 13.0% at 170 °C and 8.7% to 8.2% at 190 °C in treated RH as the reaction time increased, which corresponded to 7.5% to 7.6% and 4.8% to 4.0% on the basis of the untreated RH. Consequently, approximately 91% and 74% to 86% of the glucan was retained in the treated solids after the organosolv fractionation of RH at

170 and 190 °C, respectively, while about 80% to 86% and 92% to 97% of hemicellulose (XMG) and approximately 67% and 79% to 82% of lignin was dissolved in the fractionation liquid at 170 °C and 190 °C, respectively. The results suggested that more than two-thirds of the major components of RH can be separated in the form of a solid and liquid by applying optimized pretreatment conditions.

Table 2. Effects of the reaction conditions on the solid remaining and fractionation performances for three agricultural residues.

Biomass	Reaction Conditions		Carbohydrate and Lignin Contents			S.R. [1] (wt%)	Fractionation Results		
	Temp. (°C)	Time (min)	Glucan (wt%)	XMG [2] (wt%)	Lignin (wt%)		Glucan Retention (%)	XMG [3] Dissolved (%)	Lignin Dissolved (%)
Rice Husk	Untreated	-	35.6	13.6	22.7	100.0	-	-	-
	170	30	54.0	4.6	12.3	60.7	92.3	79.7	67.0
		60	54.1	3.9	12.3	59.4	90.5	82.8	67.8
		90	55.4	3.3	13.0	58.6	91.3	85.8	66.4
	190	30	55.6	1.9	8.7	54.9	85.9	92.4	78.9
		60	55.5	1.3	8.3	52.5	82.0	95.1	80.8
		90	54.1	0.7	8.2	48.5	73.8	97.4	82.4
Rice Straw	Untreated	-	32.9	16.2	11.7	100.0	-	-	-
	170	30	39.2	18.3	14.3	81.7	97.5	7.8	0.3
		60	40.4	16.9	13.0	75.7	93.1	20.9	15.8
		90	43.1	14.6	11.3	67.2	88.3	39.3	35.4
	190	30	46.6	14.7	11.0	67.4	95.7	38.8	36.8
		60	49.2	12.3	9.8	61.6	92.3	53.1	48.7
		90	51.8	10.2	9.0	57.7	90.8	63.5	55.9
Barley Straw	Untreated	-	41.5	18.1	14.9	100.0	-	-	-
	170	30	51.9	18.4	16.2	75.6	94.5	23.1	17.6
		60	56.6	16.9	14.7	68.3	93.1	36.3	32.7
		90	61.2	15.1	13.1	60.7	89.4	49.3	46.5
	190	30	61.2	14.2	13.0	62.6	92.5	50.9	45.1
		60	66.0	11.8	11.9	56.4	89.7	63.2	55.0
		90	70.2	9.4	10.4	51.5	87.2	73.2	63.9

[1] S.R.: Solid remaining after reaction; [2] XMG: Xylan + mannan + galactan; [3] Data in the table based on the treated biomass weight (oven dry) after reaction.

The glucan contents of fractionated RS increased from 32.9% to 39.2% to 43.1% at 170 °C and increased to 46.7% to 51.8% at 190 °C with an increasing reaction time while the XMG contents in fractionated RS decreased from 16.2% to 18.3% to 14.6% at 170 °C and to 14.7% to 10.3% at 190 °C. On the other hand, the lignin contents increased slightly from 11.7% to 14.3% to 11.2% at 170 °C, and decreased to 11.0% to 9.0% at 190 °C with an increasing reaction time. This change in the content of each component was due to the different hydrolysis reaction rates for each component during the fractionation reaction. Consequently, approximately 88% to 97% and 91% to 96% of the glucan was retained in the fractionated solids through the organosolv fractionation of RS at 170 and 190 °C, respectively, while about 8% to 39% and 39% to 64% of XMG and 0% to 35% and 37% to 56% of lignin was dissolved into the fractionation liquid at 170 °C and 190 °C, respectively. Thus, RS showed higher glucan retention, but the fractionation effect of XMG and lignin was not significant. Therefore, for RS, the organosolv fractionation process may not be an appropriate method for the effective fractionation of XMG and lignin.

In the case of BS, the glucan contents in the fractionated solid increased from 41.5% to 51.9% to 61.2% at 170 °C and increased to 61.3% to 70.2% at 190 °C. The XMG contents in fractionated BS decreased in all samples from 16.2% to 18.3% to 14.6% at 170 °C and 14.7% to 10.3% at 190 °C. However, the lignin contents decreased slightly from 18.1% to 18.4% to 15.1% at 170 °C, and to 14.2% to 9.4% at 190 °C for an increased reaction time. Approximately 89% to 95% and 87% to 92% of the glucan was retained in the fractionated BS solids at 170 °C and 190 °C, respectively, while about 23% to 49% and 51% to 73% of XMG and 18% to 47% and 45% to 64% of lignin were dissolved in the liquid hydrolysate at 170 and 190 °C, respectively. In the case of BS, glucan retention in the fractionated solid was the highest (95%) for all tested samples. In addition, the fractionation of XMG and lignin, even though it was lower than that of RH, was affected significantly by the changing reaction conditions.

3.4. Sugar Recovery and Decomposition Behavior in the Fractionated Hydrolysate

During the course of the fractionation processing, a xylose-rich liquid sample was generated by the hydrolysis of hemicellulose. In addition, usually, more severe reaction conditions make xylose undergo a condensation reaction, resulting in furfural and formic acid, which serve as strong inhibitors in the following conversion steps using enzymes and microbes [45]. To determine the characteristic behavior of sugars and other byproducts during reaction, the yields of hemicellulose-derived sugar (mainly xylose (XMG)) and byproducts, i.e., furfural and formic acid and acetic acid, were analyzed since both furfural and formic acid are the degraded products derived from hemicellulose, and they are well-known inhibitors in the bioconversion step, and acetic acid is usually formed by deacetylation of hemicellulose in the chemical fractionation reaction [45].

For RH, the hydrolysis yields of hemicellulose for various reaction times at the reaction temperature of 170 °C are shown in Figure 3. The highest hemicellulose-derived sugar (XMG) yield was 73.8% when RH was fractionated at 170 °C for 30 min, and decreased to 55.4% as the reaction time was extended to 90 min (Figure 3a). As shown in Figure 3b, a similar behavior was observed at 190 °C. However, it dramatically dropped from 67.5% at 30 min to 27.0% at 90 min, which was attributed to the decomposition of hemicellulose-derived sugar into furfural and formic acid under the harsher reaction conditions. Indeed, as the reaction time was prolonged, the formation of sugar decomposition products also increased, with furfural concentrations increasing from 1.1 to 1.9 g/L, while the formic acid concentration remained almost constant at 0.4 to 0.45 g/L at 170 °C. As seen in Figure 3b, the concentrations of furfural and formic acid increased from 2.0 to 3.5 g/L and from 0.5% to 0.8%, respectively, at 190 °C. On the other hand, the acetic acid concentration was measured to be approximately 1.2 g/L, and there was no significant change under the various reaction conditions. Levulinic acid and HMF concentrations were at relatively low levels (<0.3%).

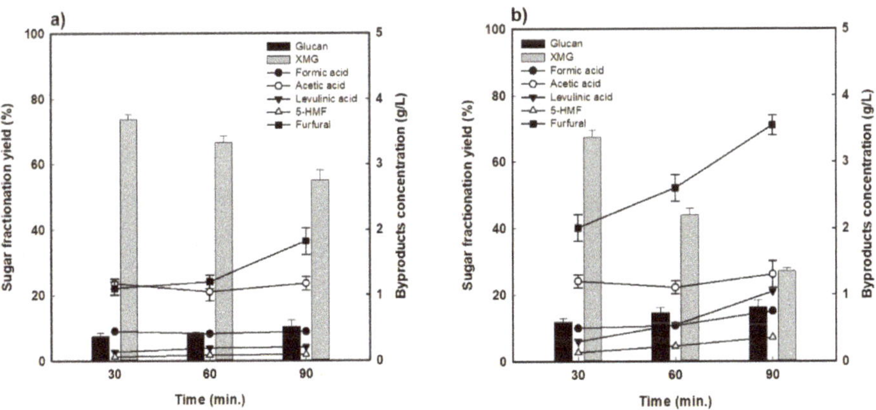

Figure 3. Hemicellulose-derived sugar yield and concentration of sugar-decomposed products in rice husk-fractionated hydrolysate (**a**) 170 °C and (**b**) 190 °C.

Figure 4 shows the sugar recovery and formation of decomposition products, such as the RS hydrolysate, as a function of the reaction time ((a) 170 °C and (b) 190 °C). The yield of XMG increased slightly over the tested reaction time to 15.2% to 31.9% at 170 °C, but the XMG yield decreased substantially from 24.4% to 17.64% at 190 °C. Therefore, in the case of RS, it was seen that the fractionation process under the tested conditions was not effective for the production of hemicellulose-derived sugar.

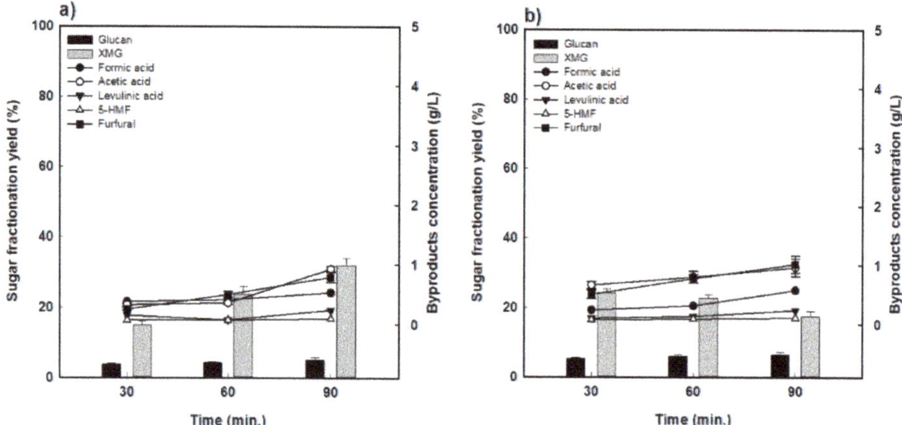

Figure 4. Hemicellulose-derived sugar yield and concentration of sugar-decomposed products in rice straw-fractionated hydrolysate (**a**) 170 °C and (**b**) 190 °C.

In the case of BS (Figure 5), the XMG yield was relatively lower than that of RH for all the reaction conditions, and the values obtained exhibited a similar trend. A higher XMG yield (approximately 35%) was achieved at 170 °C for 90 min of reaction. At the lower temperature (170 °C), the yield increased gradually as the reaction time increased from 25.2% to 35.7%. However, at the higher temperature (190 °C), the XMG yield remained constant (approximately 33%) in spite of the reaction time. The concentration of byproducts showed a distinct trend in which more sugar was decomposed under more severe reaction conditions. At 170 °C, the furfural concentration increased from 0.49 to 0.97 g/L with the increased reaction time, and the formic acid concentration also increased from 0.42 to 0.49 g/L. At higher temperatures, the furfural and formic acid concentrations increased from 0.8 to 1.8 g/L and from 0.5 to 0.70 g/L, respectively. In addition, the acetic acid concentrations showed a monotonic increase from 0.8 to 1.2 g/L at 170 °C and 0.9 to 1.3 g/L at 190 °C with an increase in the reaction time even though the hydrolytic sugar yield was not high. The lowered levels of levulinic acid (0.1–0.4 g/L) and HMF (0.1–0.2 g/L) at each tested temperature were believed to be due to lower glucose production in the fractionated hydrolyzate (Figure 5a,b).

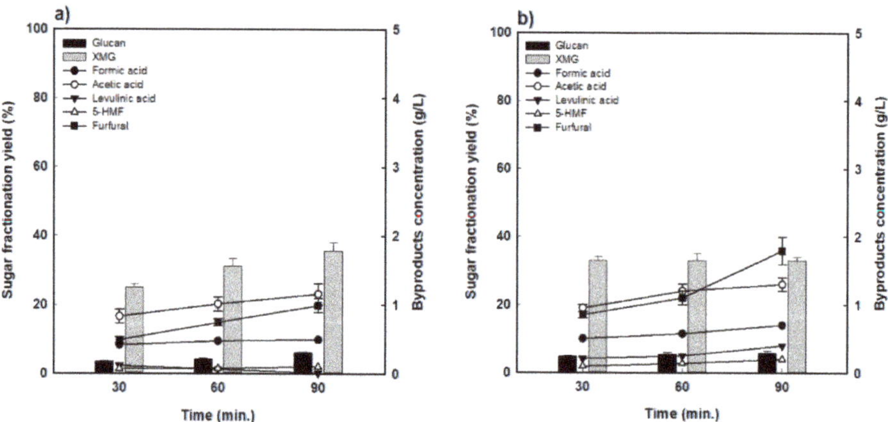

Figure 5. Hemicellulose-derived sugar yield and concentration of sugar-decomposed products in barley straw-fractionated hydrolysate (**a**) 170 °C and (**b**) 190 °C.

3.5. Enzymatic Digestibility of Fractionated Solid Residues

To evaluate the digestibility of fractionated solid for cellulose-derived sugar (glucose) production, an enzymatic hydrolysis was conducted with cellulase loading of 15 FPU/g-glucan. The enzymatic hydrolysis rates of the fractionated solid residues were compared with the reaction rates of untreated RH, RS, and BS. As shown in Figure 6, the highest digestibility (85.2% at 72 h of hydrolysis) was observed from the fractionated RH, and the highest digestibility of RS and BS was 60.0% and 70.5%, respectively, at the specific reaction condition, which resulted in the highest lignin extraction. The increase of the glucose production in enzyme-digested fractionated solid residues was higher than that of untreated RH, RS, and BS by 3.5-, 11.1-, and 3.2-fold, respectively. In the case of fractionated RS, the pretreatment effect seems to be very large despite the lowest degree of enzymatic hydrolysis. This is due to the fundamentally low enzymatic digestibility of 5.4% in untreated rice straw. On the other hand, the enzymatic digestibility of untreated RH and BS was 24.5% and 22.0%, respectively. These results suggest that the factor that greatly affects the enzyme hydrolysis reaction was the removal of components other than cellulose from the lignocellulosic biomass.

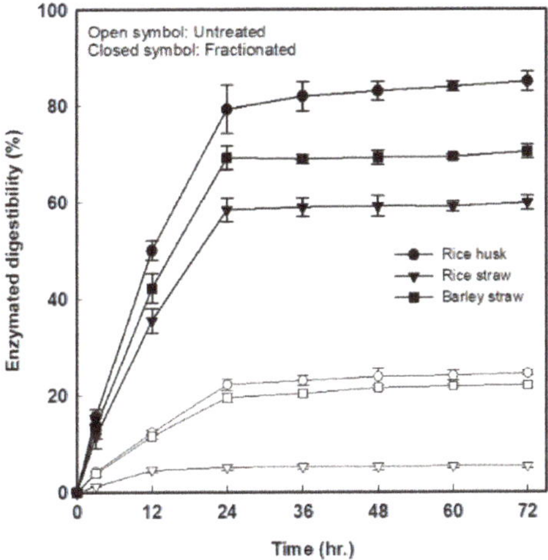

Figure 6. The 72-h enzymatic digestibility of three fractionated agricultural residues with cellulase loadings of 15 FPU/g-glucan for cellulose-derived sugar production.

3.6. Chemical Characteristics of Extracted Lignin from RS, RH, and BS

The hydroxyl groups of lignin extracted under the fractionation conditions—ethanol concentration of 60%, reaction temperature of 190 °C, and reaction time of 90 min—that showed the highest lignin extraction yields on each agricultural residues were quantified by ^{31}P NMR spectrum and are shown in Table 3, which compares the contents of p-hydroxylphenyl, guaiacyl, and syringyl units (referred to as H, G, and S, respectively) in each residue. In the RH lignin, H, G, and S contents were 0.55, 1.49, and 0.90 mmol/g, respectively, which were found to be higher than those obtained from RS or BS. In the aliphatic hydroxyl content of each extracted lignin, RH lignin showed the lowest content of 0.93 mmol/g, and the RS and BS lignins were present at 1.84 and 1.66 mmol/g, respectively. In contrast, the phenolic hydroxyl content of RH lignin had the highest value of 2.94 mmol/g and RS and BS lignin had corresponding contents of 2.19 and 1.67 mmol/g, respectively. It is generally known that the aliphatic hydroxyl can be reduced by a mechanism involving the loss of γ-methylol groups while removing formaldehyde when the β-aryl ether bonds are cleaved by ethanol [46]. In addition, according to Baucher and Sun's reports, phenolic hydroxyl units exist in biomass, forming lignin/phenolic–carbohydrate units owing to the combination of ester and ether with hemicellulose [47,48]. For this reason, the lower aliphatic hydroxyl group and higher phenolic hydroxyl group contents in the treated biomass samples may be related to the degree of hemicellulose-derived sugar release from RH.

The M_n, M_w, and polydispersity index (PDI) of extracted lignin from RS, RH, and BS were determined by GPC. In the organosolv process, as the reaction conditions become more severe, smaller lignin fragments were formed, which resulted in the decreased molecular weight of lignin. At the same fractionation conditions, the Mn of RH, RS, and BS were 1497, 1523, and 1703 g/mol, respectively. The PDI represents the molecular weight distribution of lignin, which is considered to be an important factor when lignin forms polymers [49]. As shown in Table 3, the PDI values of organosolv lignin ranged from 1.15 to 1.28, indicating a fairly uniform molecular weight distribution. Therefore, it was confirmed that organosolv-extracted lignin from three agricultural residues had high purity (more than 95%) and a low molecular weight (1453–1817 g/mol) and low PDI (1.15–1.28).

Table 3. Structural characteristics of extracted lignin fractions from three agricultural residues.

Biomass	Lignin Monomers						Molecular Weight		
	Aliphatic	H Unit [1]	G Unit [2]	S Unit [3]	Phenols	COOH	M_n	M_w	PDI [4]
	(mmol/g)	(mmol/g)	(mmol/g)	(mmol/g)	(mmol/g)	(mmol/g)	(g/mol)	(g/mol)	(-)
Rice husk	0.93	0.55	1.49	0.90	2.94	0.21	1453	1771	1.22
Rice straw	1.84	0.57	0.98	0.64	2.19	0.33	1523	1756	1.15
Barley straw	1.66	0.34	0.74	0.60	1.67	0.34	1703	2110	1.24

[1] H Unit: *p*-hydroxyphenyl; [2] G Unit: Guaiacyl; [3] S Unit: Syringyl; [4] PDI: Polydispersity index (M_w/M_n).

4. Conclusions

To address the importance of lignin valorization in the biorefinery industry, an organosolv fractionation involving ball milling was conducted at a bench scale on three agricultural residues. Lignin extracted from RH, RS, and BS showed relatively high purity (99.5% for RH and RS, and 96.8% for BS) and similar chemical characteristics. However, considering the simultaneous production of hemicellulose-derived sugars, different fractionation behaviors were shown for the three agricultural residues. Among the three tested agricultural residues, it was concluded that more effective results were obtained for RH considering its performance, including the lignin recovery (55.2%), lignin purity (99.5%), hemicellulose-derived sugar yield (73.8%), and glucan-to-glucose yield (85.2%). Fine refinements in this process can improve the fractionation efficiency of RH. Based on the results, it is speculated that the fractionation efficiency for RH can be improved with fine refinements. Furthermore, it is considered that the range of reaction conditions should be shifted for effective BS fractionation. Unfortunately, for the case of RS, other fractionation processes should be applied to achieve effective fractionation performance.

Author Contributions: T.H.K. (Tae Hoon Kim) as the first author conducted all experiments, summarized the data, and drafted the manuscript. H.K. as the co-author conducted experiments and analyzed the data. T.H.K. (Tae Hyun Kim) and K.K.O., as the co-corresponding authors, equally contributed; i.e., designed the reactor system as well as the overall study and experiments, interpreted the results, wrote the manuscript, and finalized the manuscript. All authors read and approved the final manuscript.

Funding: This research was supported by the Technology Development Program to Solve Climate Changes of the National Research Foundation (NRF) funded by the Ministry of Science and ICT (2017M1A2A2087627).

Conflicts of Interest: The authors declare no conflict of interest.

References

1. Barakat, A.; Vries, H.; Rouau, X. Dry fractionation process as an important step in current and future lignocellulose biorefineries: A review. *Bioresour. Technol.* **2013**, *134*, 362–373. [CrossRef]
2. Cherubini, F. The biorefinery concept: Using biomass instead of oil for producing energy and chemicals. *Energy Convers. Manag.* **2010**, *51*, 1412–1421. [CrossRef]

3. Matsakas, L.; Gao, Q.; Jansson, S.; Rova, U.; Christakopoulos, P. Green conversion of municipal solid wastes into fuels and chemicals. *Electron. J. Biotechnol.* **2017**, *26*, 69–83. [CrossRef]

4. Katsimpouras, C.; Zacharopoulou, M.; Matsakas, L.; Rova, U.; Christakopoulos, P.; Topakas, E. Sequential high gravity ethanol fermentation and anaerobic digestion of steam explosion and organosolv pretreated corn stover. *Bioresour. Technol.* **2017**, *244*, 1129–1136. [CrossRef]

5. Momayez, F.; Karimi, K.; Taherzadeh, M.J. Energy recovery from industrial crop wastes by dry anaerobic digestion: A review. *Ind. Crops Prod.* **2019**, *129*, 673–687. [CrossRef]

6. Statista. Available online: http://www.statista.com/statistics/255937/leading-rice-producers-worldwide/ (accessed on 16 October 2019).

7. Statista. Available online: http://www.statista.com/statistics/271973/world-barley-production-since-2008/ (accessed on 16 October 2019).

8. Kim, S.H.; Gregory, J.M. Straw to Grain Ratio Equation for Combine Simulation. *J. Biosyst. Eng.* **2015**, *40*, 314–319. [CrossRef]

9. John, R.P.; Nampoothiri, K.M.; Pandey, A. Fermentative production of lactic acid from biomass: An overview on process developments and future perspectives. *Appl. Microbiol. Biotechnol.* **2007**, *74*, 524–534. [CrossRef] [PubMed]

10. Ragauskas, A.J.; Williams, C.K.; Davison, B.H.; Britovsek, G.; Cairney, J.; Eckert, C.A.; Frederick, W.J.; Hallett, J.P.; Leak, D.J.; Liotta, C.L.; et al. The path forward for biofuels and biomaterials. *Science* **2006**, *311*, 484–489. [CrossRef] [PubMed]

11. Hahn-Hägerdal, B.; Galbe, M.; Gorwa-Grauslund, M.F.; Lidén, G.; Zacchi, G. Bio-ethanol–The fuel of tomorrow from the residues of today. *Trends Biotechnol.* **2006**, *24*, 549–556. [CrossRef]

12. Kai, D.; Tan, M.J.; Chee, P.L.; Chua, Y.K.; Yap, Y.L.; Loh, X.J. Towards lignin-based functional materials in a sustainable world. *Green Chem.* **2016**, *18*, 1175–1200. [CrossRef]

13. Kong, F.; Wang, S.; Price, J.T.; Konduri, M.K.; Fatehi, P. Water soluble kraft lignin–acrylic acid copolymer: Synthesis and characterization. *Green Chem.* **2015**, *17*, 4355–4366. [CrossRef]

14. Achinivu, E.C.; Howard, R.M.; Li, G.; Gracz, H.; Henderson, W.A. Lignin extraction from biomass with protic ionic liquids. *Green Chem.* **2014**, *16*, 1114–1119. [CrossRef]

15. Vishtal, A.G.; Kraslawski, A. Challenges in industrial applications of technical lignins. *Bio. Res.* **2011**, *6*, 3547–3568.

16. Hu, J.J.; Zhang, Q.G.; Lee, D.J. Kraft lignin biorefinery: A perspective. *Bioresour. Technol.* **2018**, *247*, 1181–1183. [CrossRef]

17. Løhre, C.; Kleinert, M.; Barth, T. Organosolv extraction of softwood combined with lignin-to-liquid-solvolysis as a semi-continuous percolation reactor. *Biomass Bioenergy* **2017**, *99*, 147–155. [CrossRef]

18. Yuan, Z.; Long, J.; Wang, T.; Shu, R.; Zhang, Q.; Ma, L. Process intensification effect of ball milling on the hydrothermal pretreatment for corn straw enzymolysis. *Energy Convers.* **2015**, *101*, 481–488. [CrossRef]

19. Castro, R.C.A.; Fonseca, B.F.; Santos, H.T.L.; Ferreira, I.S.; Mussatto, S.I.; Roberto, I.C. Alkaline deacetylation as a strategy to improve sugars recovery and ethanol production from rice straw hemicellulose and cellulose. *Ind. Crops Prod.* **2017**, *106*, 65–73. [CrossRef]

20. Mussatto, S.I. (Ed.) Biomass pretreatment with acids. In *Biomass Fractionation Technologies for a Lignocellulosic Feedstock Based Biorefinery*; Elsevier Inc.: Amsterdam, The Netherlands, 2016; pp. 169–185.

21. Barakat, A.; Chuetor, S.; Monlau, F.; Solhy, A.; Rouau, X. Eco-friendly dry chemo-mechanical pretreatments of lignocellulosic biomass: Impact on energy and yield of the enzymatic hydrolysis. *Appl. Energy* **2014**, *113*, 97–105. [CrossRef]

22. Zakaria, M.R.; Hirata, S.; Hassan, M.A. Combined pretreatment using alkaline hydrothermal and ball milling to enhance enzymatic hydrolysis of oil palm mesocarp fiber. *Bioresour. Technol.* **2014**, *169*, 236–243. [CrossRef]

23. Kim, S.M.; Dien, B.S.; Tumbleson, M.E.; Rausch, K.D.; Singh, V. Improvement of sugar yields from corn stover using sequential hot water pretreatment and disk milling. *Bioresour. Technol.* **2016**, *216*, 706–713. [CrossRef]

24. Deng, A.; Ren, J.; Wang, W.; Li, H.; Lin, Q.; Yan, Y.; Sun, R.; Liu, G. Production of xylo-sugars from corncob by oxalic acid-assisted ball milling and microwave-induced hydrothermal treatments. *Ind. Crops Prod.* **2016**, *79*, 137–145. [CrossRef]

25. Inoue, H.; Yano, S.; Endo, T.; Sakaki, T.; Sawayama, S. Combining hot-compressed water and ball milling pretreatments to improve the efficiency of the enzymatic hydrolysis of eucalyptus. *Biotechnol. Biofuels* **2008**, *1*, 2. [CrossRef] [PubMed]

26. Silva, A.S.; Inoue, H.; Endo, T.; Yano, S.; Bon, E.P.S. Milling pretreatment of sugarcane bagasse and straw for enzymatic hydrolysis and ethanol fermentation. *Bioresour. Technol.* **2010**, *101*, 7402–7409. [CrossRef] [PubMed]

27. Lin, Z.; Huang, H.; Zhang, H.; Zhang, L.; Yan, L.; Chen, J. Ball milling pretreatment of corn stover for enhancing the efficiency of enzymatic hydrolysis. *Appl. Biochem. Biotechnol.* **2010**, *162*, 1872–1880. [CrossRef] [PubMed]

28. Pan, X.; Arato, C.; Gilkes, N.; Gregg, D.; Mabee, W.; Pye, K.; Xiao, Z.; Zhang, X.; Saddler, J. Biorefining of softwoods using ethanol organosolv pulping: Preliminary evaluation of process streams for manufacture of fuel-grade ethanol and co-products. *Biotechnol. Bioeng.* **2005**, *90*, 473–481. [CrossRef]

29. Zhao, X.; Cheng, K.; Liu, D. Organosolv pretreatment of lignocellulosic biomass for enzymatic hydrolysis. *Appl. Microbiol. Biotechnol.* **2009**, *82*, 815–827. [CrossRef]

30. Zakzeski, J.; Bruijnincx, P.C.A.; Jongerius, A.L.; Weckhuysen, B.M. The catalytic valorization of lignin for the production of renewable chemicals. *Chem. Rev.* **2010**, *110*, 3552–3599. [CrossRef]

31. Torre, M.; Moral, A.; Hernández, M.; Cabeza, E.; Tijero, A. Organosolv lignin for biofuel. *Ind. Crops Prod.* **2013**, *45*, 58–63. [CrossRef]

32. Kim, S.J.; Um, B.H.; Im, D.J.; Lee, J.H.; Oh, K.K. Combined Ball Milling and Ethanol Organosolv Pretreatment to Improve the Enzymatic Digestibility of Three Types of Herbaceous Biomass. *Energies* **2018**, *11*, 2457. [CrossRef]

33. Koo, B.W.; Kim, H.Y.; Park, N.; Lee, S.M.; Yeo, H.; Choi, I.G. Organosolv pretreatment of Liriodendron tulipifera and simultaneous saccharification and fermentation for bioethanol production. *Biomass Bioenergy* **2011**, *35*, 1833–1840. [CrossRef]

34. Sluiter, A.; Ruiz, R.; Scarlata, C.; Sluiter, J.; Templeton, D. *Determination of Extractives in Biomass*; NREL/TP-510-42619; National Renewable Energy Laboratory: Golden, CO, USA, 2012.

35. Sluiter, A.; Hames, B.; Ruiz, R.; Scarlata, C.; Sluiter, J.; Templeton, D. *Determination of Structural Carbohydrates and Lignin in Biomass*; NREL/TP-510-42618; National Renewable Energy Laboratory: Golden, CO, USA, 2008.

36. Selig, M.; Weiss, N.; Ji, Y. *Enzymatic Saccharification of Lignocellulosic Biomass*; NREL/TP-510-42629; National Renewable Energy Laboratory: Golden, CO, USA, 2008.

37. Carolina, C.M.; Arturo, J.G.; Mahmoud, E.H. A comparison of pretreatment methods for bioethanol production from lignocellulosic materials. *Process Saf. Environ. Prot.* **2012**, *90*, 189–202. [CrossRef]

38. Limayem, A.; Ricke, S.C. Lignocellulosic biomass for bioethanol production: Current perspectives, potential issues and future prospects. *Prog. Energy Combust. Sci.* **2012**, *38*, 449–467. [CrossRef]

39. Menon, V.; Rao, M. Trends in bioconversion of lignocellulose: Biofuels, platform chemicals & biorefinery concept. *Prog. Energy Combust. Sci.* **2012**, *38*, 522–550. [CrossRef]

40. Mor, S.; Manchanda, C.K.; Kansal, S.K.; Ravindra, K. Nanosilica extraction from processed agricultural residue using green technology. *J. Clean. Prod.* **2017**, *143*, 1284–1290. [CrossRef]

41. Shen, Y. Rice husk silica derived nanomaterials for sustainable applications. *Renew. Sustain. Energy Rev.* **2017**, *80*, 453–466. [CrossRef]

42. Lee, J.H.; Kwon, J.H.; Lee, J.W.; Lee, H.S.; Chang, J.H.; Sang, B.I. Preparation of high purity silica originated from rice husks by chemically removing metallic impurities. *J. Ind. Eng. Chem.* **2017**, *50*, 79–85. [CrossRef]

43. Cotana, F.; Cavalaglio, G.; Nicolini, A.; Gelosia, M.; Coccia, V.; Petrozzi, A. Lignin as co-product of second generation bioethanol production from ligno-cellulosic biomass. *Energy Procedia* **2014**, *45*, 52–60. [CrossRef]

44. Mesa, L.; Gonzalez, E.; Cara, C.; González, M.; Castro, E.; Mussattoc, S.I. The effect of organosolv pretreatment variables on enzymatic hydrolysis of sugarcane bagasse. *Chem. Eng. J.* **2011**, *168*, 1157–1162. [CrossRef]

45. Lee, J.Y.; Ryu, H.J.; Oh, K.K. Acid-catalyzed hydrothermal severity on the fractionation of agricultural residues for xylose-rich hydrolyzates. *Bioresour. Technol.* **2013**, *132*, 84–90. [CrossRef]

46. Yáñez-S, M.; Matsuhiro, B.; Nuñez, C.; Pan, S.; Hubbell, C.A.; Sannigrahi, P.; Ragauskas, A.J. Physicochemical characterization of ethanol organosolv lignin (EOL) from Eucalyptus globulus: Effect of extraction conditions on the molecular structure. *Polym. Degrad. Stab.* **2014**, *110*, 184–194. [CrossRef]

47. Baucher, M.; Monties, B.; Van Montagu, M.; Boerjan, W. Biosynthesis and genetic engineering of lignin. *Crit. Rev. Plant Sci.* **1998**, *17*, 125–197. [CrossRef]

48. Sun, R.C.; Sun, X.F.; Wang, S.Q.; Zhu, W.; Wang, X.Y. Ester and ether linkages between hydroxycinnamic acids and lignins from wheat, rice, rye, and barley starws, maize stems, and fast-growing poplar wood. *Ind. Crops Prod.* **2002**, *15*, 179–188. [CrossRef]

49. McClelland, D.J.; Motagamwala, A.H.; Li, Y.; Rover, M.R.; Wittrig, A.M.; Wu, C.; Huber, G.W. Functionality and molecular weight distribution of red oak lignin before and after pyrolysis and hydrogenation. *Green Chem.* **2017**, *19*, 1378–1389. [CrossRef]

Article

Application of Sulfated Tin (IV) Oxide Solid Superacid Catalyst to Partial Coupling Reaction of α-Pinene to Produce Less Viscous High-Density Fuel

Seong-Min Cho [1], Chang-Young Hong [2], Se-Yeong Park [3], Da-Song Lee [1], June-Ho Choi [1], Bonwook Koo [4] and In-Gyu Choi [1,5,6,*]

[1]   Department of Forest Sciences, College of Agriculture and Life Sciences, Seoul National University, Seoul 08826, Korea; csmin93@snu.ac.kr (S.-M.C.); ds0429@snu.ac.kr (D.-S.L.); jhchoi1990@snu.ac.kr (J.-H.C.)

[2]   Department of Forest Biomaterials, College of Natural Resources, North Carolina State University, Raleigh, NC 27695, USA; chong6@ncsu.edu

[3]   Department of Forest Biomaterials Engineering, College of Forest and Environment Science, Kangwon National University, Chuncheon 24341, Korea; parksy319@kangwon.jac.kr

[4]   Intelligent & Sustainable Materials R&D Group, Korea Institute of Industrial Technology (KITECH), Cheonan 31056, Korea; bkoo@kitech.re.kr

[5]   Research Institute of Agriculture and Life Sciences, College of Agriculture and Life Sciences, Seoul National University, Seoul 08826, Korea

[6]   Institutes of Green-Bio Science and Technology, Seoul National University, Pyeongchang 25354, Korea

*   Correspondence: cingyu@snu.ac.kr; Tel.: +82-2-880-4785

Received: 19 April 2019; Accepted: 16 May 2019; Published: 18 May 2019

Abstract: Brønsted acid-catalyzed reactions of α-pinene have been studied because of their ability to produce various types of fragrance molecules. Beyond this application, dimeric hydrocarbon products produced from coupling reactions of α-pinene have been suggested as renewable high-density fuel molecules. In this context, this paper presents the application of a sulfated tin(IV) oxide catalyst for the partial coupling reaction of α-pinene from turpentine. Brønsted acid sites inherent in this solid superacid catalyst calcined at 550 °C successfully catalyzed the reaction, giving the largest yield of dimeric products (49.6%) at 120 °C over a reaction time of 4 h. Given that the low-temperature viscosity of the mentioned dimeric products is too high for their use as a fuel in transportation engines, lowering the viscosity is an important avenue of study. Therefore, our partial coupling reaction of α-pinene provides a possible solution as a considerable amount of the isomers of α-pinene still remained after the reaction, which reduces the low-temperature viscosity. On the basis of a comparison of the reaction products, a plausible mechanism for the reaction involving coinstantaneous isomerization and coupling reaction of α-pinene was elucidated.

Keywords: solid superacid catalyst; sulfated tin(IV) oxide; α-pinene partial coupling; renewable high-density fuel

---

1. Introduction

Turpentine, one of the most widely produced plant-derived secondary metabolites, is a mixture of monoterpenes. It is made up mainly of α-pinene and its isomers, such as β-pinene and camphene. Depending on the production method, it is categorized as gum turpentine (produced from oleoresins of conifers), wood turpentine (produced from aged pine stumps), sulfate turpentine (produced by the kraft pulping process), and sulfite turpentine (produced by the sulfite pulping process) [1]. Concerns about fossil fuel depletion and environmental destruction urge us to develop alternative energy resources. In this regard, turpentine, in which C10 hydrocarbons form major components, has been reported as a potential candidate for providing biofuel blends for fueling both spark ignition engines and compression ignition engines [2–5].

A coupling reaction by which C20 hydrocarbons can be synthesized from renewable α-pinene has been devised for ramjet propulsion [6–8], but not for conventional jet fuels. This is because the relatively high carbon number of these compared to that of petroleum-derived fuels tremendously increases their low-temperature viscosity, which limits the suitability of using dimeric products alone for transport fuel [9]. Thus, the partial coupling of monoterpene hydrocarbons was suggested as one possible solution [10]. This agrees with the results reported by Harvey et al., who suggested blending dimeric products with monoterpene hydrocarbons such as α-pinene, thereby resolving the viscosity problem of dimeric products [6].

Coupling of monoterpene hydrocarbons has been studied using various types of heterogeneous acid catalysts, such as Nafion, Nafion SAC-13, montmorillonite K10, Al-incorporated MCM-41, Pd-Al-incorporated MCM-41, phosphotungstic acid supported on MCM-41, phosphotungstic acid supported on SiO_2, Zeolite Hβ, and silica-alumina aerosol [6–8,11–14]. Given that the isomerization of monoterpene hydrocarbons is usually attributed to Brønsted acid activity and that Brønsted acids can also assist in their coupling reactions undertaken in harsh conditions, it has been thought that the catalytic activities of the above catalysts are caused by Brønsted acid sites and the role of Lewis acid sites is less significant.

Herein, we used sulfated tin(IV) oxide (SO_4^{2-}/SnO_2) as a solid superacid catalyst to carry out partial coupling of α-pinene for the production of less viscous high-density fuel molecules. SO_4^{2-}/SnO_2 catalysts have been used in various types of organic reactions, such as dehydration of sorbitol and xylose to isosorbide and furfural, respectively, esterification of free fatty acids, the Penchmann condensation reaction, and deprotection of silyl ether groups [15–19]. Successive chemical precipitation and immersion in diluted sulfuric acid yielded this catalyst from tin chloride pentahydrate in a facile procedure. In this study, the catalyst successfully furnished dimeric products from α-pinene with its isomers in solventless conditions. On the basis of the results, a plausible mechanism for the isomerization and coupling reaction, in which Brønsted acid catalysis plays a central role, was also suggested. To our knowledge, no attempt has been made so far to propose a mechanism considering both reactions together.

2. Materials and Methods

2.1. Catalyst Preparation

To prepare the catalysts, tin oxide powder obtained from tin (IV) chloride pentahydrate ($SnCl_4 \cdot 5H_2O$) by chemical precipitation, followed by drying, was used as a precursor. Briefly, a transparent 0.1 M tin chloride solution was prepared by dissolving $SnCl_4 \cdot 5H_2O$ in distilled water. To hydrolyze the tin chloride complex, a 28 wt% aqueous ammonia solution was added dropwise under vigorous stirring. The addition was stopped when the pH of the solution reached approx. 8. After a precipitate was separated from clear supernatant liquid, thorough washing was carried out with a 4 wt% ammonium acetate solution by centrifugation. The white product was then dried in an oven at 105 °C for 24 h and ground into a fine powder. The prepared tin oxide (SnO_2) powder (10 g) was placed in a round flask containing 3 g of sulfuric acid diluted with 20 mL of distilled water. After sufficient stirring at 80 °C, water was removed in vacuo and sulfuric acid-treated SnO_2 was dried and stored in an oven at 65 °C. This precatalyst was dried further at 120 °C for 12 h, followed by calcination at 450, 550, 600, and 650 °C for 4 h.

2.2. Catalyst Characterization

To understand the properties of the prepared catalysts and elucidate their catalytic activity, we conducted X-ray diffraction (XRD) analysis, field-emission scanning electron microscopy (FE-SEM), thermogravimetric analysis (TGA), and temperature-programmed desorption of ammonia (NH_3-TPD).

The XRD patterns of the catalysts were collected to compare the crystal structures. The surface morphology of the catalysts was investigated using FE-SEM. Energy dispersive X-ray spectroscopy (EDS) and elemental mapping were also performed as part of the SEM investigation. To study the thermal properties of the catalysts, TGA was carried out. NH_3-TPD was performed to characterize the improvement in the catalyst acidity. The detailed analysis conditions are described in the Supplementary Materials.

2.3. Catalytic Tests

In a typical experiment, turpentine (5 g), tridecane (2 g, GC internal standard), and catalysts (0.1 g) were added without any solvent to a 50 mL glass flask equipped with a magnetic Teflon-coated stirrer and a reflux condenser. The reactor was then loaded on a preheated aluminum heating block and stirred vigorously. Upon completion of the reaction, the reactor was removed from the heating block and immediately cooled to room temperature using a cold-water bath. After cooling, the crude reaction mixture was diluted with n-hexane (100 mL) and filtered over a Celite pad. To estimate the product composition by the internal standard method, a diluted reaction mixture was analyzed using a gas chromatograph (7890B) equipped with a DB-5ms column (30 m × 250 μm, 0.25 μm thickness), a mass spectrometer detector (5977A), and a flame ionization detector (Agilent Technologies, Santa Clara, CA, USA). The yield of the products and the conversion rate of α-pinene were calculated using the following equations:

$$\text{Yield of product } i \ (\%) \ = \ \frac{\text{Weight of product } i}{\text{Weight of crude reaction mixture}} \times 100, \qquad (1)$$

$$\text{Conversion rate of } \alpha-\text{pinene } (\%) \ = \ \frac{\text{Consumed weight of } \alpha-\text{pinene}}{\text{Initial weight of } \alpha-\text{pinene}} \times 100. \qquad (2)$$

3. Results and Discussion

3.1. Catalyst Characteristics

3.1.1. Catalyst Surface Morphology

To investigate the surface morphological properties of intact tin oxide (SnO_2) and sulfated tin oxide (SO_4^{2-}/SnO_2), field emission-scanning electron microscopy (FE-SEM) observation was carried out. The SEM images in Figure 1a,b show that intact SnO_2 has a rough surface, whereas SO_4^{2-}/SnO_2 exhibits a smooth one because, although both surfaces consist of globular nanoparticles, the size of the nanoparticles in SO_4^{2-}/SnO_2 was much smaller than that of the nanoparticles in intact SnO_2. Energy dispersive spectroscopy (EDS) analysis confirmed the presence of sulfur-containing groups on the surface of SO_4^{2-}/SnO_2 (Figure S1a–d). In addition, elemental mapping images clearly show a uniform distribution of sulfur atoms on the surface (Figure 1c and Figure S1e).

3.1.2. Catalyst Crystal Structure

The crystal structures of tin oxide (SnO_2) and sulfated tin oxide (SO_4^{2-}/SnO_2) calcined at 550 °C was characterized using powder X-ray diffraction, as shown in Figure 2. There were obvious differences in the diffraction peaks for the species. The diffractogram of intact SnO_2 shows a good match to the ICDD powder diffraction file of cassiterite SnO_2 (PDF 00-041-1445), which means that it has a tetragonal crystal structure similar to that of cassiterite SnO_2 (P42/mnm space group). Major diffraction peaks observed at $2\theta = 27°$, $34°$, and $52°$ could be indexed to the (110), (101), and (211) planes of cassiterite SnO_2, respectively. Immersion in diluted sulfuric acid before calcination significantly influenced the crystal structure. Even though the characteristic diffraction peaks of cassiterite SnO_2 were detected in the diffractogram of SO_4^{2-}/SnO_2, the intensities of the peaks decreased significantly and their breadth broadened considerably. Because this weakening and broadening denotes diminished crystallinity and crystallite size [16,20], Figure 2 suggests an amorphous structure and small-sized crystallites of SO_4^{2-}/SnO_2. In accordance with the results presented in previous studies related to sulfated metal oxides such as SnO_2 and ZrO_2, sulfate groups on the surface of sulfuric acid immersed SnO_2 seem to hamper both aggregation and crystallization themselves during calcination [17,21,22]. This result also coincides with the FE-SEM images shown in Figure 1, which shows the difference in the size of globular nanoparticles making up the surfaces.

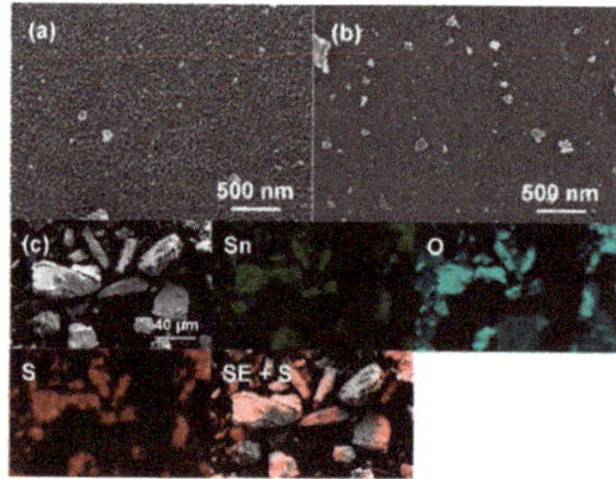

Figure 1. Field-emission scanning electron microscopy (FE-SEM) images of the surface of (**a**) intact SnO_2; and (**b**) SO_4^{2-}/SnO_2; (**c**) SEM-energy dispersive X-ray spectroscopy (EDS) elemental mapping images of SO_4^{2-}/SnO_2. The samples were prepared by calcination at 550 °C for 4 h. In the overlay image of SEM and sulfur atom mapping, the gray color represents an area in shadow or surface facing away from the EDS detector.

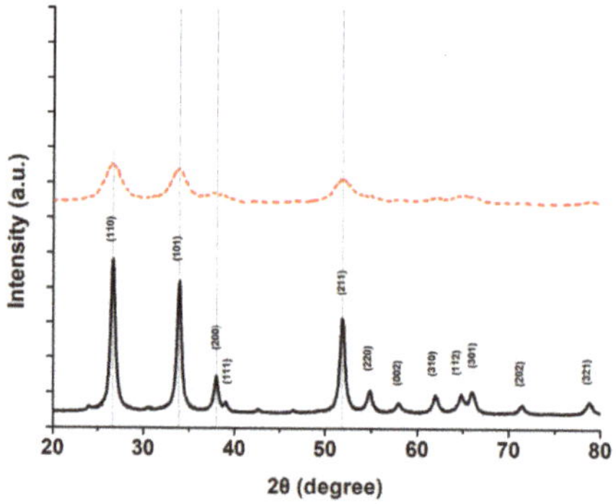

Figure 2. X-ray diffraction patterns of intact SnO_2 (black line) and SO_4^{2-}/SnO_2 (red dashed line). The samples were calcined at 550 °C. On the basis of the ICDD powder diffraction file (PDF 00-041-1445), the peaks were indexed as tetragonal crystal structure as cassiterite SnO_2 ($P4_2/mnm$ space group).

3.1.3. Catalyst Thermostability

The catalyst was designed to be applied to the coupling of α-pinene under somewhat harsh reaction conditions (≥100 °C). Our query was whether the catalytic activity over sulfated tin oxide (SO_4^{2-}/SnO_2) could be sustained during the reaction. Because sulfate groups introduced on the surface of the SO_4^{2-}/SnO_2 support by immersion in diluted sulfuric acid are responsible for the catalytic activity [23], the thermal decomposition behavior of sulfate groups was investigated by thermogravimetric analysis (TGA). Figure S2 shows the TGA graph of intact SnO_2 and SO_4^{2-}/SnO_2 calcined at 550 °C. SO_4^{2-}/SnO_2 displayed two distinguishable weight loss sections, whereas intact SnO_2 presented gradual weight reduction throughout the temperature range. The weight reduction of intact SnO_2 and the first weight loss of SO_4^{2-}/SnO_2 from 120 to 550 °C can be explained by the desorption of chemisorbed water molecules and dehydroxylation on the surface of the SnO_2 support [20,22]. The difference between the extents of the weight reduction experienced by them may be attributed to the differences in the surface functional groups, especially sulfate groups, on which water can interact [23]. In addition, significant weight loss was observed at temperatures higher than 600 °C in the case of SO_4^{2-}/SnO_2, which is attributed to the decomposition of sulfate groups on the surface of the SnO_2 support [20,22].

3.1.4. Catalyst Acidity

To qualitatively evaluate the acidities of the intact tin oxide (SnO_2) and sulfated tin oxide (SO_4^{2-}/SnO_2) catalysts, we carried out temperature programmed desorption of ammonia (NH_3-TPD). The TPD profiles of desorbed ammonia clearly show an improvement in the acid strength of the catalysts when SnO_2 was immersed in diluted sulfuric acid before calcination (Figure 3a). Peak deconvolution can help us interpret the meaning of the overlapped peaks. As can be seen in Figure 3b, the profile of intact SnO_2 calcined at 550 °C consisted mainly of two peaks centered at approx. 372 °C (dark yellow) and 513 °C (dark cyan), which are attributed to weak and strong chemisorption of ammonia. On the other hand, the SO_4^{2-}/SnO_2 calcined at 550 °C had four kinds of acid sites which present four peaks centered at approx. 240 °C (magenta), 353 °C (blue), 531 °C (purple), and 736 °C (orange) (Figure 3c). In particular, the strongest acid sites were attributed to the desorption of ammonia from sulfate groups on the surface of SO_4^{2-}/SnO_2 [24]. This interpretation coincided with the TGA results for SO_4^{2-}/SnO_2 shown in Figure S2, which displays a significant weight loss at temperatures higher than 600 °C due to the decomposition of sulfate groups on the surface. In addition to the obvious differences in the peak number and overall peak intensities in the TPD profiles, the amount of NH_3-uptake by the catalysts also suggests that immersion in diluted sulfuric acid generates much more acid sites than those developed in intact SnO_2 (Table 1).

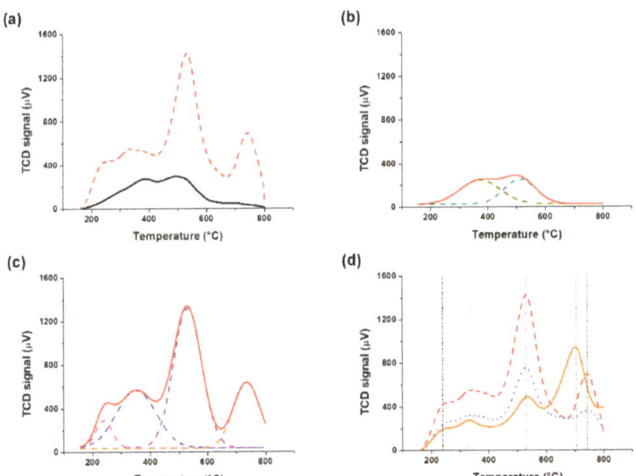

Figure 3. (a) Temperature programmed desorption of ammonia (NH_3–TPD) profiles of intact SnO_2 (black line) and SO_4^{2-}/SnO_2 (red dashed line); (**b–c**) Deconvolution of overlapped peaks in the profiles. The samples were prepared by calcination at 550 °C for 4 h; (**d**) NH_3–TPD profiles of SO_4^{2-}/SnO_2 calcined at 550 °C (red dashed line), 600 °C (blue dotted line), and 650 °C (orange line).

The effect of the calcination temperature on the acidity of the sulfated catalyst was evaluated based on the amount of NH_3 uptake (Table 1). Generally, increasing the calcination temperature reduces the total acid sites of sulfated or phosphated solid acid catalysts [25,26]. Our results coincide with those reported. However, one difference was found above the 600 °C calcination temperature, with a new peak centered at approx. 696 °C appearing and the disappearance of the peak centered at approx. 736 °C in the TPD profiles (Figure 3d).

Table 1. NH_3 uptake of intact SnO_2 and $SO_4{}^{2-}/SnO_2$ calcined at different temperatures.

Catalyst	Intact SnO_2 [1]	$SO_4{}^{2-}/SnO_2$			
		450 °C [2]	550 °C	600 °C	650 °C
NH_3 uptake (μmol g^{-1})	171.1 [3]	1161.8	760.2	400.1	425.0

[1] Intact SnO_2 was calcined at 550 °C. [2] Calcination temperature. [3] Calculated based on the NH_3–TPD results.

3.2. Catalytic Tests and Reaction Mechanism

3.2.1. Effect of Catalyst Calcination Temperature on Partial Coupling Reaction of α-Pinene

When it comes to preparing a sulfated metal oxide catalyst using sulfuric acid treatment, a sintering process is important because the promotion of sulfate groups on the surface of metal oxide occurs during calcination and the acid sites generated by these sulfate groups offer the essential catalytic activity [23]. Figure 4 shows stark differences in the effect of the calcination temperature on the catalytic activity of sulfated tin oxide ($SO_4{}^{2-}/SnO_2$). The yield of dimeric products from α-pinene (**1**) was the largest for the catalyst calcined at 550 °C (48.9 ± 1.2%), while almost no conversion to dimeric products and significantly lowered production were observed below and above this temperature, respectively. When we tried to correlate the yield of dimeric products with the amount of the total acid sites of catalysts, unlike previous papers [25,26], there was a large discrepancy in the calcination temperature range from 450 to 550 °C (Table 1 and Figure 4). Although the catalyst calcined at 450 °C showed the highest NH_3 uptake, it was not able to furnish dimeric products at all. Since a large amount of sulfuric acid can hinder the growth of SnO_2 crystals (Figure 2), thereby leading to poor promotion of sulfate groups on the surface of $SO_4{}^{2-}/SnO_2$, the negligible catalytic activity of catalyst calcined at 450 °C (α-pinene conversion rate <20%) was attributed to sulfuric acid remaining in large quantities during the calcination process. Zhang et al. reported that increasing the calcination temperature from 150 to 550 °C made sulfate groups of $SO_4{}^{2-}/CeO_2$ transition from surface sulfate states to bulk sulfate states; the catalyst mainly possessing sulfate groups as surface sulfate states worked well in catalytic reduction of NO by NH_3. According to the Raman spectra in which only the catalyst calcined at 550 °C presented the peaks denoting bulk sulfate states, this transition seems to occur abruptly when the calcination temperature increased from 450 °C to 550 °C [27]. In this respect, the discrepancy between the yield of dimeric products and the amount of the total acid sites of catalysts can be understood by considering that the catalyst possessing sulfate groups as bulk sulfate states is effective in the coupling reaction of α-pinene. On the other hand, calcination temperatures higher than 550 °C caused significant decomposition of sulfuric acid and even sulfate groups on the surface of $SO_4{}^{2-}/SnO_2$ (Figure S2). Therefore, when the same amount of sulfuric acid was treated, a much higher calcination temperature made the catalyst lose sulfate groups. In this regard, the less effective catalytic activity of catalysts calcined at 600 and 650 °C can be understood. Finally, to compare the effect of the promotion of sulfate groups on the catalytic activity, we conducted the reaction with intact SnO_2 calcined at 550 °C, which showed negligible α-pinene conversion (data not shown).

In contrast to dimeric products, catalysts prepared at higher calcination temperatures (600 and 650 °C) more readily furnished camphene (**6**) from α-pinene (**1**) and the catalyst calcined at 550 °C also yielded a considerable amount of compound **6** (Figure 4). In other words, the conversion to compound **6** competed with the production of dimeric products from compound **1**, and once formed, compound **6** was thought to be indifferent to the homocoupling reaction with these catalysts. This is supported by the fact that several catalysts lack the ability to catalyze homocoupling of compound **6** to form dimeric hydrocarbon products [11].

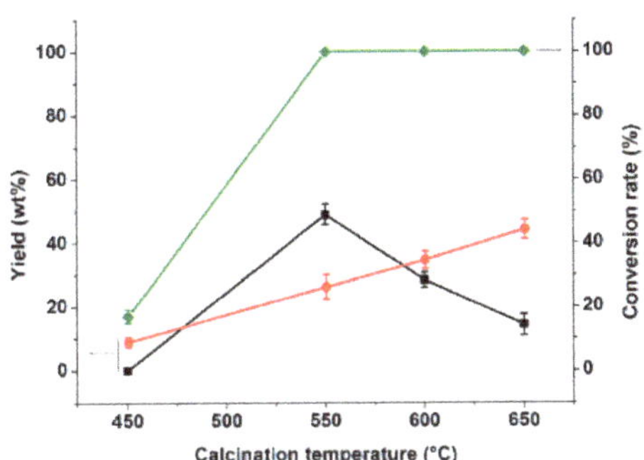

Figure 4. Effect of catalyst calcination temperature on the yields of dimeric products (■) and camphene (●) and the conversion rate of α-pinene (♦). All the other reaction conditions were maintained for 3 h at 120 °C.

3.2.2. Effect of Reaction Time and Temperature on Partial Coupling Reaction of α-Pinene

To investigate the effects of reaction time and temperature on the yield of dimeric products, we selected the sulfated tin oxide (SO_4^{2-}/SnO_2) catalyst calcined at 550 °C based on the results shown in Figure 4. As one can see in Figure 5, the conversion of α-pinene (**1**) was almost 100 ± 0% only after 30 min except for the reaction at 100 °C. This consumption furnished dimeric products almost entirely at the beginning of the reaction (Figure 5a). The additional increase in the yield continued up to 3 h at all reaction temperatures. A quite interesting feature was that after 3 h, there was no significant difference among the yields of dimeric products at 100, 110, and 130 °C, although the increase in the yield of dimeric products during the reaction was the highest at 100 °C (from 13.6 ± 3.2% to 45.0 ± 0.7%). A significant difference after 3 h was only observed at 120 °C, which produced the highest yield of all (48.9 ± 1.2%). However, when we carried out the reaction further in these conditions, there was a poor improvement (from 48.9 ± 1.2% to 49.6 ± 0.7%). In other words, only around half of compound **1** yielded dimeric products and therefore our catalyst did not seem suitable for the coupling reaction. However, considering the low-temperature viscosity of dimeric products [9], this partial coupling will be even more appropriate for fuel applications [10].

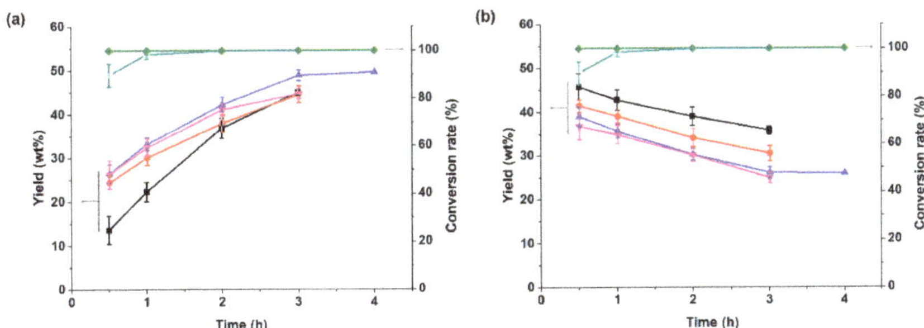

Figure 5. Effect of reaction time and temperature on the yields of (**a**) dimeric products and (**b**) camphene: (■) for 100 °C; (●) for 110 °C; (▲) for 120 °C; (▼) for 130 °C; (■) for α-pinene conversion rate except for 100 °C condition; (+) for α-pinene conversion rate at 100 °C condition. The used catalyst was prepared by calcination at 550 °C for 4 h.

Figure 5b shows that the yields of camphene (**6**) decreased depending on the reaction temperature and time. Given compound **6** seems to be less reactive to homocoupling with our catalyst, its decrease during the reaction time implies the possibility of heterocoupling of compound **6** with other species, even including dimeric products. Because such heterocoupling can consume dimeric products and furnish oligomeric products, reaction temperatures higher than 120 °C produced a lower yield of dimeric products.

In order to investigate why the yields of both dimeric products and camphene did not show major changes after 3 h, we conducted a reusable test of the catalyst to prove whether or not the deactivation of the catalyst occurred during the reaction. Unfortunately, the used catalyst did not furnish dimeric products (data not shown). We thought that this was due to the sludge generated by the condensation of various substrates covering the acid sites of the catalyst. An interesting result was that thorough washing with acetone also cannot restore the activity of the used catalyst, but rather eliminated the activity. This is because the acetone washing removed not only the sludge but also the sulfate groups of bulk sulfate states [27]. As mentioned before, sulfate groups as bulk sulfate states of catalysts seem to have key role in the coupling of α-pinene (**1**).

3.2.3. Mechanism of Isomerization of α-Pinene over Sulfated Tin Oxide

Finally, the mechanism of isomerization and coupling reaction of α-pinene (**1**) over the sulfated tin oxide (SO_4^{2-}/SnO_2) catalyst was considered. The generally accepted mechanism of acid-catalyzed isomerization of compound **1** includes two distinct pathways [28]: ring enlargement rearrangement (path **A**), wherein tri- or bicyclic compounds are produced; and ring opening rearrangement (path **B**), in which monocyclic compounds are formed (Scheme 1). Both pathways are initiated by protonation of olefin in compound **1** generating pinanyl cation (**2**). If the highly strained 4-membered ring in carbocation **2** suffers from ring enlargement by Wagner–Meerwein rearrangement, bornanyl cation (**3**) is formed. This carbocation **3** can produce tricyclene (**4**) via direct deprotonation or camphene (**6**) via deprotonation of isocamphanyl cation (**5**) resulting from 1,2-sigmatropic rearrangement of carbocation **3**. Alternatively, carbocation **2** can be stabilized by the opening of the four-membered ring, which results in *p*-mentha-1-en-8-yl cation (**7**). In light of the similarity in structure and stability between carbocation **7** and *p*-mentha-1-en-4-yl cation (**8**), the 1,2-hydride shift between them is considered reversible. These carbocations (**7** and **8**) can be deprotonated to furnish monocyclic compounds (**9–12**).

In the above mechanism, the selectivities of camphene (**6**) from path **A** and limonene (**9**) from path **B** are of interest, and it is undeniable that the selectivities depend on what catalyst is used. Because compound **9** can further isomerize to other monocyclic compounds, catalysts showing the higher (but not by much) selectivity for compound **6** compared to that for compound **9** have been reported more commonly. Kitano et al. reported isomerization of α-pinene (**1**) over an Al_2O_3-supported MoO_3 catalyst, which presented a slightly higher selectivity toward compound **6** than for compound **9**, although the conversion of compound **1** was not good enough [29]. SiO_2- or MCM-41-supported $H_3PW_{12}O_{40}$ and MSU-S BEA or Y catalysts showed over 90% conversion of compound **1** and a slightly higher selectivity for compound **6** [12,30,31]. Furthermore, very powerful catalysts such as Fe^{3+}-exchanged clinoptilolite, sulfated zirconia, SBA-15 supported TiO_2, and sulfuric acid-treated montmorillonite clay have been suggested for the production of compound **6** with prominent selectivity [28,32–35]. However, catalysts that showed the higher selectivity for compound **9** compared to compound **6** have been reported much less frequently. Yamamoto et al. developed a SiO_2-supported Pr_2O_3 catalyst, which showed very high selectivity for compound **9** although the conversion of compound **1** was notably low [36]. In addition to this catalyst, a SiO_2-supported $AlCl_3$ catalyst showed higher (but not by much) selectivity for compound **9** with the varying conversion of compound **1**. In this study, SO_4^{2-}/SnO_2 showed much higher selectivity for compound **6** than for compound **9** with 100% conversion of compound **1** (Figure 5b and Figure S3). This tendency can be justified by the difference between compounds **6** and **9** in reactivity for further isomerization as mentioned previously. It has also been reported that Al-MCM-41 lacks the ability to catalyze homocoupling of compound **6** formed by the isomerization of compound **1** [11]. Given the significant amount of compound **6** still remaining after the coupling reaction was over, not only further isomerization but also homocoupling of compound **6** seems to be difficult with our catalyst, as mentioned in the previous section.

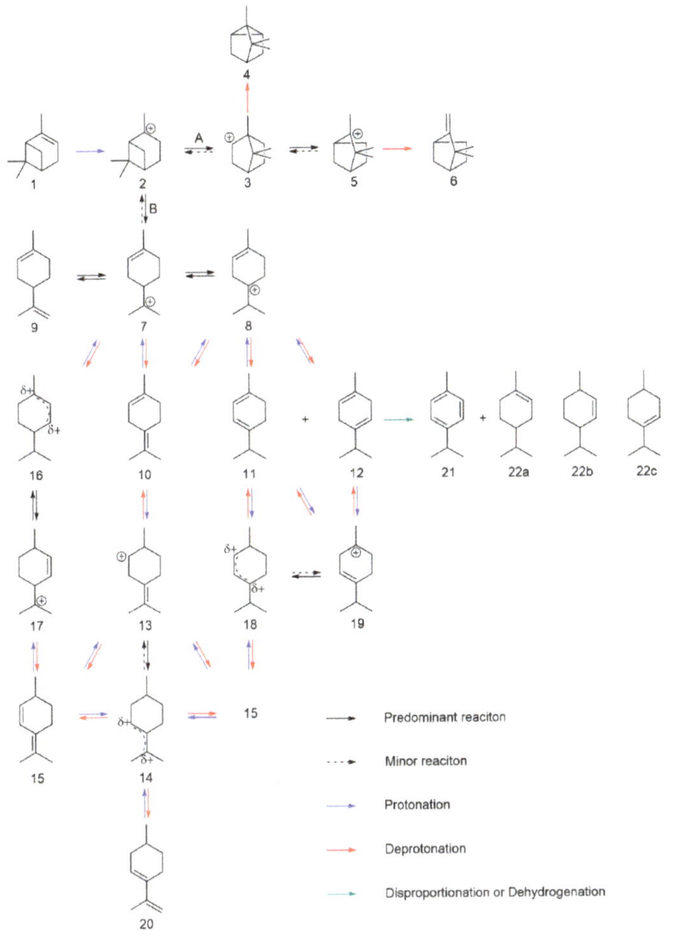

Scheme 1. Plausible mechanism of isomerization of α-pinene over sulfated tin oxide catalyst: α-pinene (**1**); pinanyl cation (**2**); bornanyl cation (**3**); tricyclene (**4**); isocamphanyl cation (**5**); camphene (**6**); *p*-mentha-1-en-8-yl cation (**7**); *p*-mentha-1-en-4-yl cation (**8**); limonene (**9**); terpinolene (**10**); α-terpinene (**11**); γ-terpinene (**12**); *p*-mentha-4(8)-en-2-yl cation (**13**); allylic carbocation I (**14**); isoterpinolene (**15**); allylic carbocation II (**16**); *p*-mentha-2-en-8-yl cation (**17**); allylic carbocation III (**18**); *p*-mentha-3-en-1-yl cation (**19**); *p*-mentha-3,8-diene (**20**); *p*-cymene (**21**); *p*-menthene isomers (**22a–c**).

In some papers, isoterpinolene (**15**) has been also suggested as a co-product. One plausible mechanism for the formation of compound **15** starts from protonation of terpinolene (**10**), resulting in *p*-mentha-4(8)-en-2-yl cation (**13**) [28]. Not only can direct deprotonation of carbocation **13** generate compound **15**, but also deprotonation of allylic carbocation I (**14**) resulting from carbocation **13** via 1,2-hydride shift can do the same. However, this suggestion has been controversial considering various isoterpinolene/terpinolene concentration ratios, either higher or lower than 1 [37]. A second possible mechanism is initiated by protonation of α-terpinene (**11**) or γ-terpinene (**12**), which gives allylic carbocation III (**18**) and *p*-mentha-3-en-1-yl cation (**19**). This proposal was supported by Salacinski's results which showed the chemical equilibria of *p*-menthadiene species under sulfuric acid at 67 °C [38]. When compound **11** or **12** reacted under this condition as a sole starting material, the chemical equilibrium consisted of only compounds **11**, **12**, **15**, and a small amount of *p*-mentha-3,8-diene

(**20**), where stabilization by the formation of a conjugated diene was considered the driving force. In addition, the author described reaction coordinate diagrams with allylic carbocations (**14** and **18**) as reaction intermediates. The presence of compound **20** in our results, even though the quantity of it was relatively small, also seems to indicate this mechanism. Additionally, the successive transformation of *p*-mentha-1-en-8-yl cation (**7**) into allylic carbocation II (**16**) and *p*-mentha-2-en-8-yl cation (**14**) via 1,3- and 1,5-hydride shifts, respectively, was suggested as the other possible route for the formation of compound **15** [39]. Behr and Wintzer also reported that compound **15** was formed as a major side product when the hydroaminomethylation of compound **9** was carried out with a [Rh(cod)Cl]$_2$/TPPTS catalyst [40], which means that compound **9** can be a linchpin when it comes to the production of compound **15**. As seen in Figure S3, the yield of compound **15** seems to follow the same trend as that of **9** along the reaction time. This also suggests that the third suggested pathway chiefly occurs during the isomerization of α-pinene (**1**) in the case of our catalyst.

In addition to the above-isomerized products, *p*-cymene (**21**) and *p*-menthene isomers (**22a–c**) were detected in the reaction mixture. The simultaneous formation of compounds **21** and **22a–c** can be explained by disproportionation between α-terpinene (**11**) and γ-terpinene (**12**) [6,11]. Moreover, dehydrogenation of compounds **11** and **12** was suggested to justify the production of compound **21** with the generation of hydrogen gas [41,42]. Given the para position of methyl and isopropyl groups therein, compound **21** was generally reported as the target product not only when α-pinene (**1**) was used in neat form [41,43] or as a major constituent of crude sulfate turpentine [44], but also when limonene (**9**) was used as a sole starting material [42,45]. The concentration of compound **21** in the reaction mixture gradually increased with reaction time (Figure S3). This is because it did not participate in further reactions, including both the isomerization and coupling reaction in our catalytic system [7,12].

3.2.4. Mechanism of Coupling of α-Pinene over Sulfated Tin Oxide

The lack of knowledge about the molecular structure of dimeric products obtained from monoterpene is attributed to the simultaneous homo- and heterocoupling that occurs for the starting materials and the isomers therein [6,11,12]. Furthermore, it being hard to isolate only one dimeric product from a product mixture, the study of their molecular structure has proven difficult. These phenomena were also observed in our results; although the reaction started with α-pinene (**1**) as a sole substance, besides isomerization, a variety of dimeric products were concurrently produced (Figure S4). Nevertheless, some reports have suggested several possible molecular structures without an understanding of the complicated reaction system [7,11,46].

Acid-catalyzed coupling reactions of monoterpene hydrocarbons generally involve three steps: first, the protonation of olefin in monoterpene giving carbocations; next, the attack of olefin (nucleophile) in another monoterpene on the previous carbocation (electrophile) furnishing a dimeric carbocation with the formation of a new C–C bond between the nucleophile and the electrophile; and, finally, the deprotonation of this carbocation giving dimeric products. Of course, dimeric carbocations formed by nucleophilic attack on monomeric carbocations, or even by the re-protonation of dimeric products, can suffer from isomerization and further coupling reactions, which is one reason for the complexity of the reaction mechanism. In addition to coupling reactions that involve protonation/deprotonation, the Diels–Alder reaction between monoterpenes, especially α-terpinene (**11**), has also been suggested as a possible mechanism for the coupling reaction of monoterpenes [47].

In light of the results obtained for a reaction at a relatively low temperature (100 °C, Figure S3a), although the concentration of α-pinene (**1**) in the reaction mixture precipitously decreased with reaction time, considering almost all of the dimeric products were yielded in just 30 min, compound dimeric structure (**PD1**). We can also imagine that further isomerization of this possible dimeric product gives, for example, **PD1a** and **PD1b** via ring enlargement and ring opening, respectively.

The isomers of α-pinene (**1**), such as camphene (**6**), limonene (**9**), terpinolene (**10**), α-terpinene (**11**), γ-terpinene (**12**), and isoterpinolene (**15**), and *p*-mentha-3,8-diene (**20**), all of which have olefin in their structure, can partake in coupling reactions as compound **1** does. As electrophiles, stable allylic carbocations (**14**, **16**, and **18**) derived from these monocyclic isomers were thought to play a pivotal role in the coupling reaction [38,39]. This type of coupling can occur from the beginning to the end of a reaction, particularly in a predominant coupling reaction that occurs after compound **1** is completely consumed. In Scheme 2, we suggest some possible dimeric structures (**PD2–4**), showing that the isomers react as nucleophiles and electrophiles (carbocations).

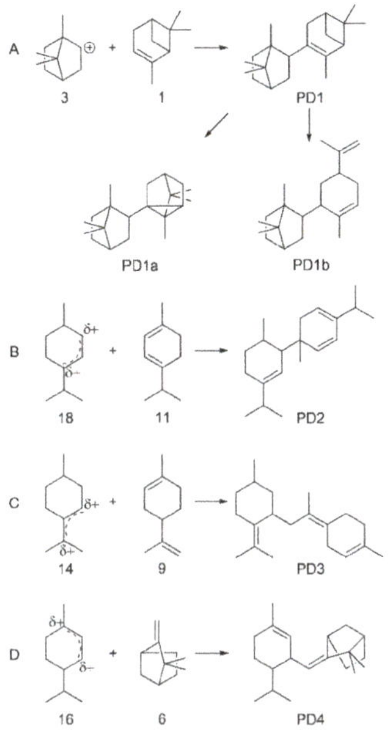

Scheme 2. Examples of possible dimeric products of the coupling reaction of α-pinene catalyzed using sulfated tin oxide: (**A**) coupling of bornanyl cation (**3**) and α-pinene (**1**) giving possible dimeric product (**PD1**) and its isomers (**PD1a–b**); (**B**) coupling of allylic carbocation III (**18**) and α-terpinene (**11**) giving **PD2**; (**C**) coupling of allylic carbocation I (**14**) and limonene (**9**) giving **PD3**; (**D**) coupling of allylic carbocation II (**16**) and camphene (**6**) giving **PD4**.

Although we are not sure whether camphene (**6**) reacts as a nucleophile or electrophile (as isocamphanyl cation (**5**)) during the coupling reaction, it is clear that compound **6** predominantly participates in a heterocoupling (Scheme 2, **PD4**) rather than a homocoupling reaction considering that a significant amount of compound **6** still remained after the reaction. This is in agreement with the results obtained using Al-incorporated MCM-41 [11]. Meylemans et al. asserted that this phenomenon is attributed to the low basicity of compound **6**, which causes poor interactions between the external olefin and the acid sites of the catalyst, thereby making the protonation of compound **6** difficult [7].

4. Conclusions

In summary, a sulfated tin(IV) oxide catalyst prepared using a facile procedure was applied to the partial coupling reaction of α-pinene to furnish a renewable and less viscous high-density fuel. To evaluate the catalytic activity of the catalyst, we considered the effect of the calcination temperature, reaction time, and reaction temperature, and attempted to rationalize the results using the catalyst characteristics. The catalyst calcination temperature had an enormous influence on the production of dimeric hydrocarbon products, while reaction times and temperatures exceeding 1 h and 100 °C affected the reaction to a lesser extent. The highest yield of dimeric products (49.6%) was obtained when the catalyst was calcined at 550 °C and the reaction was carried out at 120 °C for 4 h. Although the yield was less than half, we think this value is enough to consider utilizing the reaction products as renewable fuels, because it is not known that the dimeric products alone have a low-temperature viscosity too high for use as a fuel in transportation engines. In other words, the mixture with the isomers of α-pinene can drag down the low-temperature viscosity to the range of transportation fuels. Finally, we described the possible mechanism of the coinstantaneous isomerization and coupling reaction of α-pinene owing to our catalyst acting as a Brønsted acid.

Supplementary Materials: The following are available online at http://www.mdpi.com/1996-1073/12/10/1905/s1, Figure S1: (a,b) Energy dispersive spectroscopy (EDS) spectra, (c,d) elemental quantitative data obtained from intact SnO_2 and $SO_4{}^{2-}/SnO_2$ and (e) higher resolution of elemental mapping image of $SO_4{}^{2-}/SnO_2$, Figure S2: TGA curves of intact SnO_2 (black line) and $SO_4{}^{2-}/SnO_2$ (red dashed line), Figure S3: The yields of the monomeric products along the reaction time at (a) 100 °C and (b) 110 °C, Figure S4: General chromatogram of dimeric products extracted from GC/FID result.

Author Contributions: Conceptualization, S.-M.C. and D.-S.L.; Data curation, S.-M.C., C.-Y.H. and B.K.; Formal analysis, S.-M.C., C.-Y.H. and B.K.; Methodology, S.-M.C. and D.-S.L.; Supervision, I.-G.C.; Writing—original draft, S.-M.C., S.-Y.P. and J.-H.C.; Writing—review & editing, C.-Y.H.

Funding: This study was supported by Mid-career Researcher Program in Basic Research of National Research Foundation of Korea grant funded by the Korea government (MSIP) (NRF-2016R1A2B4014222).

Conflicts of Interest: The authors declare no conflict of interest.

References

1. Gscheidmeier, M.; Fleig, H. Turpentines. In *Ullmann's Encyclopedia of Industrial Chemistry*; BASF Aktiengesellschaft: Ludwigshafen, Germany, 2012.
2. Yumrutaş, R.; Alma, M.H.; Özcan, H.; Kaşka, Ö. Investigation of Purified Sulfate Turpentine on Engine Performance and Exhaust Emission. *Fuel* **2008**, *87*, 252–259. [CrossRef]
3. Arpa, O.; Yumrutas, R.; Alma, M. Effects of Turpentine and Gasoline-Like Fuel Obtained from Waste Lubrication Oil on Engine Performance and Exhaust Emission. *Energy* **2010**, *35*, 3603–3613. [CrossRef]
4. Anand, B.P.; Saravanan, C.; Srinivasan, C.A. Performance and Exhaust Emission of Turpentine Oil Powered Direct Injection Diesel Engine. *Renew. Energy* **2010**, *35*, 1179–1184. [CrossRef]
5. Dubey, P.; Gupta, R. Influences of Dual Bio-Fuel (Jatropha Biodiesel and Turpentine Oil) on Single Cylinder Variable Compression Ratio Diesel Engine. *Renew. Energy* **2018**, *115*, 1294–1302.
6. Harvey, B.G.; Wright, M.E.; Quintana, R.L. High-Density Renewable Fuels Based on the Selective Dimerization of Pinenes. *Energy Fuels* **2009**, *24*, 267–273. [CrossRef]
7. Meylemans, H.A.; Quintana, R.L.; Harvey, B.G. Efficient Conversion of Pure and Mixed Terpene Feedstocks to High Density Fuels. *Fuel* **2012**, *97*, 560–568.
8. Jung, J.K.; Lee, Y.; Choi, J.-W.; Jae, J.; Ha, J.-M.; Suh, D.J.; Choi, J.; Lee, K.-Y. Production of High-Energy-Density Fuels by Catalytic β-Pinene Dimerization: Effects of the Catalyst Surface Acidity and Pore Width on Selective Dimer Production. *Energy Convers. Manag.* **2016**, *116*, 72–79. [CrossRef]
9. Meylemans, H.A.; Baldwin, L.C.; Harvey, B.G. Low-Temperature Properties of Renewable High-Density Fuel Blends. *Energy Fuels* **2013**, *27*, 883–888. [CrossRef]
10. Cho, S.-M.; Kim, J.-H.; Kim, S.-H.; Park, S.-Y.; Kim, J.-C.; Choi, I.-G. A Comparative Study on the Fuel Properties of Biodiesel from Woody Essential Oil Depending on Terpene Composition. *Fuel* **2018**, *218*, 375–384. [CrossRef]
11. Zou, J.J.; Chang, N.; Zhang, X.; Wang, L. Isomerization and Dimerization of Pinene using Al-Incorporated MCM-41 Mesoporous Materials. *ChemCatChem* **2012**, *4*, 1289–1297. [CrossRef]
12. Nie, G.; Zou, J.-J.; Feng, R.; Zhang, X.; Wang, L. HPW/MCM-41 Catalyzed Isomerization and Dimerization of Pure Pinene and Crude Turpentine. *Catal. Today* **2014**, *234*, 271–277. [CrossRef]
13. Zhang, S.; Xu, C.; Zhai, G.; Zhao, M.; Xian, M.; Jia, Y.; Yu, Z.; Liu, F.; Jian, F.; Sun, W. Bifunctional Catalyst Pd–Al-MCM-41 for Efficient Dimerization–Hydrogenation of β-Pinene in One Pot. *RSC Adv.* **2017**, *7*, 47539–47546. [CrossRef]
14. Andrea Merino, N.; Cecilia Avila, M.; Alejandra Comelli, N.; Natalia Ponzi, E.; Isabel Ponzi, M. Dimerization of α-Pinene, Using Phosphotungstic Acid Supported on SiO2 as Catalyst. *Curr. Catal.* **2014**, *3*, 240–243. [CrossRef]

15. Dabbawala, A.A.; Mishra, D.K.; Hwang, J.-S. Sulfated Tin Oxide as an Efficient Solid Acid Catalyst for Liquid Phase Selective Dehydration of Sorbitol to Isosorbide. *Catal. Commun.* **2013**, *42*, 1–5. [CrossRef]

16. Suzuki, T.; Yokoi, T.; Otomo, R.; Kondo, J.N.; Tatsumi, T. Dehydration of Xylose Over Sulfated Tin Oxide Catalyst: Influences of the Preparation Conditions on the Structural Properties and Catalytic Performance. *Appl. Catal. A Gen.* **2011**, *408*, 117–124. [CrossRef]

17. Moreno, J.; Jaimes, R.; Gómez, R.; Niño-Gómez, M. Evaluation of Sulfated Tin Oxides in the Esterification Reaction of Free Fatty Acids. *Catal. Today* **2011**, *172*, 34–40. [CrossRef]

18. Ahmed, A.I.; El-Hakam, S.; Khder, A.; El-Yazeed, W.A. Nanostructure Sulfated Tin Oxide as an Efficient Catalyst for the Preparation of 7-Hydroxy-4-Methyl Coumarin by Pechmann Condensation Reaction. *J. Mol. Catal. A Chem.* **2013**, *366*, 99–108. [CrossRef]

19. Bhure, M.H.; Kumar, I.; Natu, A.D.; Rode, C.V. Facile and Highly Selective Deprotection of Tert-Butyldimethyl Silyl Ethers Using Sulfated SnO2 as a Solid Catalyst. *Synth. Commun.* **2008**, *38*, 346–353. [CrossRef]

20. Gutierrez-Baez, R.; Toledo-Antonio, J.; Cortes-Jacome, M.; Sebastian, P.; Vázquez, A. Effects of the SO4 Groups on the Textural Properties and Local Order Deformation of SnO2 Rutile Structure. *Langmuir* **2004**, *20*, 4265–4271. [CrossRef]

21. Khder, A.; El-Sharkawy, E.; El-Hakam, S.; Ahmed, A. Surface Characterization and Catalytic Activity of Sulfated Tin Oxide Catalyst. *Catal. Commun.* **2008**, *9*, 769–777. [CrossRef]

22. Qi, X.; Watanabe, M.; Aida, T.M.; Smith, R.L., Jr. Sulfated Zirconia as a Solid Acid Catalyst for the Dehydration of Fructose to 5-Hydroxymethylfurfural. *Catal. Commun.* **2009**, *10*, 1771–1775. [CrossRef]

23. Trickett, C.A.; Popp, T.M.O.; Su, J.; Yan, C.; Weisberg, J.; Huq, A.; Urban, P.; Jiang, J.; Kalmutzki, M.J.; Liu, Q. Identification of the Strong Brønsted Acid Site in a Metal–Organic Framework Solid Acid Catalyst. *Nat. Chem.* **2019**, *11*, 170–176. [CrossRef] [PubMed]

24. Kassaye, S.; Pagar, C.; Pant, K.K.; Jain, S.; Gupta, R. Depolymerization of Microcrystalline Cellulose to Value Added Chemicals Using Sulfate Ion Promoted Zirconia Catalyst. *Bioresour. Technol.* **2016**, *220*, 394–400. [CrossRef] [PubMed]

25. Qiu, L.; Wang, Y.; Pang, D.; Ouyang, F.; Zhang, C. SO42−–Mn–Co–Ce Supported on TiO2/SiO2 with High Sulfur Durability for Low-Temperature SCR of NO with NH3. *Catal. Commun.* **2016**, *78*, 22–25. [CrossRef]

26. Saravanan, K.; Park, K.S.; Jeon, S.; Bae, J.W. Aqueous Phase Synthesis of 5-Hydroxymethylfurfural from Glucose over Large Pore Mesoporous Zirconium Phosphates: Effect of Calcination Temperature. *ACS Omega* **2018**, *3*, 808–820. [CrossRef]

27. Zhang, L.; Zou, W.; Ma, K.; Cao, Y.; Xiong, Y.; Wu, S.; Tang, C.; Gao, F.; Dong, L. Sulfated Temperature Effects on the Catalytic Activity of CeO2 in NH3-Selective Catalytic Reduction Conditions. *J. Phys. Chem. C* **2015**, *119*, 1155–1163. [CrossRef]

28. Yadav, M.K.; Chudasama, C.D.; Jasra, R.V. Isomerisation of α-Pinene Using Modified Montmorillonite Clays. *J. Mol. Catal. A Chem.* **2004**, *216*, 51–59. [CrossRef]

29. Kitano, T.; Okazaki, S.; Shishido, T.; Teramura, K.; Tanaka, T. Brønsted Acid Generation of Alumina-Supported Molybdenum Oxide Calcined at High Temperatures: Characterization by Acid-Catalyzed Reactions and Spectroscopic Methods. *J. Mol. Catal. A Chem.* **2013**, *371*, 21–28. [CrossRef]

30. Da Silva Rocha, K.A.; Robles-Dutenhefner, P.A.; Kozhevnikov, I.V.; Gusevskaya, E.V. Phosphotungstic Heteropoly Acid as Efficient Heterogeneous Catalyst for Solvent-Free Isomerization of α-Pinene and Longifolene. *Appl. Catal. A Gen.* **2009**, *352*, 188–192. [CrossRef]

31. Wang, J.; Hua, W.; Yue, Y.; Gao, Z. MSU-S Mesoporous Materials: An Efficient Catalyst for Isomerization of α-Pinene. *Bioresour. Technol.* **2010**, *101*, 7224–7230. [CrossRef]

32. Akgül, M.; Özyağcı, B.; Karabakan, A. Evaluation of Fe-And Cr-Containing Clinoptilolite Catalysts for the Production of Camphene from α-Pinene. *J. Ind. Eng. Chem.* **2013**, *19*, 240–249. [CrossRef]

33. Comelli, N.A.; Ponzi, E.N.; Ponzi, M.I. α-Pinene Isomerization to Camphene: Effect of Thermal Treatment on Sulfated Zirconia. *Chem. Eng. J.* **2006**, *117*, 93–99. [CrossRef]

34. Comelli, N.A.; Ponzi, E.N.; Ponzi, M.I. Isomerization of α-Pinene, Limonene, α-Terpinene, and Terpinolene on Sulfated Zirconia. *J. Am. Oil Chem. Soc.* **2005**, *82*, 531–535. [CrossRef]

35. Wróblewska, A.; Miądlicki, P.; Makuch, E. The Isomerization of α-Pinene over the Ti-SBA-15 Catalyst—The Influence of Catalyst Content and Temperature. *React. Kinet. Mech. Catal.* **2016**, *119*, 641–654. [CrossRef]

36. Yamamoto, T.; Matsuyama, T.; Tanaka, T.; Funabiki, T.; Yoshida, S. Generation of Acid Sites on Silica-Supported Rare Earth Oxide Catalysts: Structural Characterization and Catalysis for α-Pinene Isomerization. *Phys. Chem. Chem. Phys.* **1999**, *1*, 2841–2849. [CrossRef]

37. Lopes, C.; Lourenco, J.; Pereira, C.; Marcelo-Curto, M. Aromatization of Limonene with Zeolites Y. In *Natural Products in the New Millennium: Prospects and Industrial Application*; Springer: Berlin, Germany, 2002; pp. 429–436.

38. Salacinski, E.J. *Acid-Catalyzed Isomerization of the p-Menthadienes*; The University of Arizona: Tucson, AZ, USA, 1966.

39. McCormick, J.; Barton, D.L. Studies in 85% H3PO4—II: On the Role of the α-Terpinyl Cation in Cyclic Monoterpene Genesis. *Tetrahedron* **1978**, *34*, 325–330. [CrossRef]

40. Behr, A.; Wintzer, A. Hydroaminomethylation of the Renewable Limonene with Ammonia in an Aqueous Biphasic Solvent System. *Chem. Eng. Technol.* **2015**, *38*, 2299–2304. [CrossRef]

41. Al-Wadaani, F.; Kozhevnikova, E.F.; Kozhevnikov, I.V. Zn (II)–Cr (III) Mixed Oxide as Efficient Bifunctional Catalyst for Dehydroisomerisation of α-Pinene to P-Cymene. *Appl. Catal. A Gen.* **2009**, *363*, 153–156. [CrossRef]

42. Zhao, C.; Gan, W.; Fan, X.; Cai, Z.; Dyson, P.J.; Kou, Y. Aqueous-Phase Biphasic Dehydroaromatization of Bio-Derived Limonene into P-Cymene by Soluble Pd Nanocluster Catalysts. *J. Catal.* **2008**, *254*, 244–250. [CrossRef]

43. Golets, M.; Ajaikumar, S.; Mohln, M.; Wärnå, J.; Rakesh, S.; Mikkola, J.-P. Continuous Production of the Renewable ρ-Cymene from α-Pinene. *J. Catal.* **2013**, *307*, 305–315. [CrossRef]

44. Linnekoski, J.A.; Asikainen, M.; Heikkinen, H.; Kaila, R.K.; Räsänen, J.; Laitinen, A.; Harlin, A. Production of P-Cymene from Crude Sulphate Turpentine with Commercial Zeolite Catalyst Using a Continuous Fixed Bed Reactor. *Org. Process Res. Dev.* **2014**, *18*, 1468–1475. [CrossRef]

45. Martin-Luengo, M.; Yates, M.; Rojo, E.S.; Arribas, D.H.; Aguilar, D.; Hitzky, E.R. Sustainable P-Cymene and Hydrogen from Limonene. *Appl. Catal. A Gen.* **2010**, *387*, 141–146. [CrossRef]

46. Arias-Ugarte, R.; Wekesa, F.S.; Schunemann, S.; Findlater, M. Iron (III)-Catalyzed Dimerization of Cycloolefins: Synthesis of High-Density Fuel Candidates. *Energy Fuels* **2015**, *29*, 8162–8167. [CrossRef]

47. Fernandes, C.; Catrinescu, C.; Castilho, P.; Russo, P.; Carrott, M.; Breen, C. Catalytic Conversion of Limonene over ACID ACTIVATED SERRA de Dentro (SD) Bentonite. *Appl. Catal. A Gen.* **2007**, *318*, 108–120. [CrossRef]

Article

Green Diesel Production over Nickel-Alumina Nanostructured Catalysts Promoted by Copper

Mantha Gousi [1], Eleana Kordouli [1,2], Kyriakos Bourikas [2,*], Emmanouil Symianakis [3], Spyros Ladas [3], Christos Kordulis [1,2,4] and Alexis Lycourghiotis [1]

[1]   Department of Chemistry, University of Patras, GR-26504 Patras, Greece; mgousi_2@hotmail.com (M.G.); ekordouli@upatras.gr (E.K.); kordulis@upatras.gr (C.K.); alycour@upatras.gr (A.L.)

[2]   School of Science and Technology, Hellenic Open University, Tsamadou 13-15, GR-26222 Patras, Greece

[3]   Surface Science Laboratory, Department of Chemical Engineering, University of Patras, GR-26504 Patras, Greece; manos_simianakis@yahoo.com (E.S.); ladas@chemeng.upatras.gr (S.L.)

[4]   Foundation for Research and Technology, Institute of Chemical Engineering Science (FORTH/ICE-HT), Stadiou str., Platani, P.O. Box 1414, GR-26500 Patras, Greece

*   Correspondence: bourikas@eap.gr

Received: 15 June 2020; Accepted: 15 July 2020; Published: 18 July 2020

Abstract: A series of nickel–alumina catalysts promoted by copper containing 1, 2, and 5 wt. % Cu and 59, 58, and 55 wt. % Ni, respectively, (symbols: 59Ni1CuAl, 58Ni2CuAl, 55Ni5CuAl) and a non-promoted catalyst containing 60 wt. % Ni (symbol: 60NiAl) were prepared following a one-step co-precipitation method. They were characterized using various techniques (N_2 sorption isotherms, XRD, SEM-EDX, XPS, H_2-TPR, NH_3-TPD) and evaluated in the selective deoxygenation of sunflower oil using a semi-batch reactor (310 °C, 40 bar of hydrogen, 96 mL/min hydrogen flow rate, and 100 mL/1 g reactant to catalyst ratio). The severe control of the co-precipitation procedure and the direct reduction (without previous calcination) of precursor samples resulted in mesoporous nano-structured catalysts (most of the pores in the range 3–5 nm) exhibiting a high surface area (192–285 $m^2\,g^{-1}$). The promoting action of copper is demonstrated for the first time for catalysts with a very small Cu/Ni weight ratio (0.02–0.09). The effect is more pronounced in the catalyst with the medium copper content (58Ni2CuAl) where a 17.2% increase of green diesel content in the liquid products has been achieved with respect to the non-promoted catalyst. The copper promoting action was attributed to the increase in the nickel dispersion as well as to the formation of a Ni-Cu alloy being very rich in nickel. A portion of the Ni-Cu alloy nanoparticles is covered by Ni^0 and Cu^0 nanoparticles in the 59Ni1CuAl and 55Ni5CuAl catalysts, respectively. The maximum promoting action observed in the 58Ni2CuAl catalyst was attributed to the finding that, in this catalyst, there is no considerable masking of the Ni-Cu alloy by Ni^0 or Cu^0. The relatively low performance of the 55Ni5CuAl catalyst with respect to the other promoted catalysts was attributed, in addition to the partial coverage of Ni-Cu alloy by Cu^0, to the remarkably low weak/moderate acidity and relatively high strong acidity exhibited by this catalyst. The former favors selective deoxygenation whereas the latter favors coke formation. Copper addition does not affect the selective-deoxygenation reactions network, which proceeds predominantly via the dehydration-decarbonylation route over all the catalysts studied.

Keywords: green diesel; renewable diesel; Ni catalyst; biofuel; hydrodeoxygenation; Cu-promotion effect

---

1. Introduction

The emission of carbon dioxide from the combustion of fossil fuels (carbon, oil, and natural gas) is responsible for the global warming whereas the increasing demands for fossil fuels is expected to lead to their progressive depletion during the 21st century [1]. The replacement of fossil fuels by renewable ones is actually critical for facing such a double problem [1–3]. Undoubtedly, biomass is an important source of renewable energy. The animal fat and plant oils triglyceride biomass is very attractive because it is much less complex than lignocellulosic biomass whereas the ratio of oxygen to the combustible carbon and hydrogen atoms in a molecule of triglyceride is relatively small. The catalytic transformation of triglycerides into n-alkanes in the diesel range (green diesel) was initially studied over noble metals and conventional NiMo or CoMo/γ-Al$_2$O$_3$ sulfide catalysts [4]. Nickel non-sulphided catalysts have gained much interest in the last decade [5–16]. The previously mentioned transformation over nickel non-sulphided catalysts is realized by hydrotreatment (temperature: 240–360 °C, hydrogen pressure: 10–80 bar) whereas the relevant chemistry, stated in the next paragraph, is now well established.

The removal of oxygen without or with very small fragmentation extent of the side chains of triglycerides is called Selective DeOxygenation (SDO). The first step of the SDO network is the rapid hydrogenation of olefinic bonds of side chains of triglycerides. It is followed by the gradual decomposition of the O-C bonds in the glycerol side resulting progressively to di-glycerides and then mono-glycerides and then to free fatty acids and propane. The free fatty acids could be transformed by direct decarboxylation (deCO$_2$) into n-alkanes with an odd number of carbon atoms (mainly n-C17 and n-C15) and carbon dioxide. However, this pathway, if any, has very low probability above nickel catalysts [4]. Much more probable is the second pathway whereby the free fatty acids are reduced to the corresponding fatty aldehydes by water removal. The aldehydes are then decarbonylated, which results in n-alkanes with an odd number of carbon atoms (mainly n-C17 and n-C15) and carbon monoxide. Thus, the SDO via this route takes place by dehydration–decarbonylation (deH$_2$O-deCO). The third pathway involves the reduction of the fatty aldehydes into the corresponding fatty alcohols and their very rapid equilibration. The fatty alcohols are transformed into olefins by dehydration, which are then hydrogenated and leads to n-alkanes with an even number of carbon atoms (mainly n-C18 and n-C16). Thus, the SDO through this route takes place by dehydration (deH$_2$O). In parallel, the fatty alcohols may react with the fatty acids producing long chain esters [6]. These may undergo SDO resulting in hydrocarbons. The above-described SDO network is schematized in Figure S1 (Supplementary Material). In addition to the previously mentioned liquid phase reactions, reactions may take place between CO$_2$, CO, and hydrogen deliberated in the gas phase (reverse water gas shift and methanation).

Recently, we have contributed to this subject by developing a co-precipitation methodology, which ensures severe control of the precipitation parameters. This methodology allowed preparing nickel–alumina catalysts with a high specific surface area, even at high nickel loading [6]. These catalysts are very active in the SDO of sunflower oil (SFO) under solvent-free conditions and a very high SFO volume/catalyst ratio (100 mL/1 g). This work showed that the yield of green diesel over such catalysts is a linear function of the nickel surface exposed. The highest value of the latter and, thus, the maximum catalytic performance was achieved over the catalyst with nickel content at about 60 wt. % in which a good compromise between the specific surface area and nickel loading has been achieved. This preparation methodology was then successfully applied to nickel–zirconia catalysts [7] and to nickel–alumina catalysts promoted by molybdenum [17] and zinc [18]. Concerning the promoted catalysts, the total amount of the nickel content was about 60 wt. % whereas the promoters exhibited their optimum effect at very small loadings (1–6%) compared to the nickel loading (54–59%).

With the idea to extend our study to the copper promoted catalysts, we are surveying the relevant literature. The promoting action of copper concerning nickel catalysts used in the SDO of plant oils, biodiesel, residual fatty raw materials, and related compounds into green diesel has gained the research interest in the last years [19–30]. Yakovlev's group was pioneer in this subject [19–22]. They first studied Ni monometallic and Ni-Cu bimetallic catalysts supported on CeO$_2$, ZrO$_2$, or CeO$_2$-ZrO$_2$ in the SDO of biodiesel [19,20] and then a bimetallic Ni-Cu catalyst supported on γ-Al$_2$O$_3$ in the SDO of methyl palmitate and ethyl caprate [21]. The bimetallic catalysts were proved more attractive due to their efficiency to prevent methane formation. The Ni–Cu/CeO$_2$–ZrO$_2$ catalyst exhibited the highest performance partly attributed to the presence of a Ni_1-xCu_x (x = 0.2–0.3) solid solution as a constituent of the active center and partly to the development of a mixed cerium-zirconium oxide phase. The formation of the previously mentioned nickel–copper solid solution was also found in the $Ni-Cu/\gamma-Al_2O_3$ catalyst [21]. The Crocker's group has also contributed in the domain [23–26]. In the first work [23], a series of $Ni/\gamma-Al_2O_3$ catalysts were prepared containing 20 wt. % Ni and 0, 1, 2, and 5 wt. % Cu and studied in the SDO of tristearin, stearic acid, and triolein. The maximum performance was obtained over the catalyst 20% Ni–5% $Cu/\gamma-Al_2O_3$. The promoting action of copper was attributed to the increase in the surface of metallic nickel as well as to the suppression of surface coking and, hence, catalyst deactivation. This may reflect the ability of Cu to curb the cracking activity of nickel expressed in the non-promoted catalyst. This was also confirmed in the subsequent articles reported by the group where the un-promoted 20% $Ni/\gamma-Al_2O_3$ catalyst was compared to the promoted 20% Ni–5% $Cu/\gamma-Al_2O_3$ one in the SDO of residual fatty raw materials (yellow grease and hemp seed oil [24], waste free fatty acids, and brown grease [25]). They also compared the 20% Ni–5% $Cu/\gamma-Al_2O_3$ catalyst to the 20% Ni–5% $Fe/\gamma-Al_2O_3$ and 20% Ni-0.5%$Pt/\gamma-Al_2O_3$ ones in the SDO of waste cooking oil (WCO) [26]. A work reported by Jing et al. deals with the copper promoting action in $Ni/\gamma-Al_2O_3$ catalysts containing 20 wt. % Ni and 0, 1, 3, 6, or 10 wt. % Cu in the SDO of biodiesel [27]. The maximum performance was obtained over the catalyst containing 6 wt. % Ni. The decrease in catalytic performance as the copper loading increases from 6 to 10 wt. %, which was attributed to its accumulation on the support surface that leads to pore blocking. As inferred in the first work of the Crocker's group [23], the promoting action of copper was mainly attributed to coking suppression. Two interesting works were reported by the group of Fu in which methanol was used as a hydrogen donor instead of gas hydrogen [28,29]. In the first work [28], copper (30 wt. %), nickel (30 wt. %), and nickel–copper catalysts (15, 30, 60 wt. %, and Cu/Ni weight ratio = $\frac{1}{2}$) supported on ZrO_2 were studied in the SDO of oleic acid. It was reported that the formation of a Cu-Ni alloy favors SDO and inhibits cracking, which increases the catalytic performance. In the second work [29], copper (60 wt. %), nickel (60 wt. %), copper (40 wt. %)-nickel (20 wt. %), copper (30 wt. %)-nickel (30 wt. %), and copper (20 wt. %)-nickel (40 wt. %) catalysts supported on alumina were studied in the SDO of oleic acid. Cracking of C–C bonds deduced by the presence of cracked paraffins was found over the nickel monometallic catalyst. The formation of the Cu–Ni alloy in the bimetallic catalysts and the presence of partially oxidized copper favor SDO and inhibit the cracking of the C–C bonds, which leads to enhanced catalytic performance. A very recent work reported by Miao et al. [30] deals with two monometallic (10 wt. % $Ni/\gamma-Al_2O_3$, 10 wt. % $Cu/\gamma-Al_2O_3$) and five bimetallic $NiCu/\gamma-Al_2O_3$ catalysts with Cu/Ni ratios equal to 1/9, 3/7, 5/5, 7/3, and 9/1 used in the SDO of methyl laurate. They concluded that the oxide precursors can be effectively reduced at 420 °C for 2 h into the corresponding metallic catalysts, which is comprised of Ni^0, Cu^0, and NiCu alloy-supported species. The formation of the NiCu alloy promotes the electronic interactions between Ni and Cu, which enhances catalytic performance. The catalyst with the Cu/Ni ratio equal to 7/3 was proved to be the most active. The nickel and the copper active centers favor, respectively, SDO through dehydration–decarbonylation and dehydration.

In conclusion, the promoting action of copper has been demonstrated. Some aspects of the promoting action start to emerge. The copper promoting action is expressed through the increase in the nickel dispersion and the formation of nickel-copper alloy with better catalytic behavior than nickel. It seems that this curbs the C-C cracking activity of nickel, which depresses the formation of cracked paraffins, methane, and carbon deposition and favors the formation of diesel range n-alkanes and catalyst stability.

In the present work, we are continuing our research effort on co-precipitated $Ni-Al_2O_3$ catalysts by studying the copper and promoting action in these catalysts concerning the SDO of SFO. Four catalysts containing 60, 59, 58, and 55 wt. % Ni and 0, 1, 2, and 5 wt. % Cu, denoted by 60NiAl, 59Ni1CuAl, 58Ni2CuAl, and 55Ni5CuAl, were synthesized following the previously mentioned rigorous co-precipitation methodology. The catalysts were characterized using various methods and evaluated in the SDO of SFO in a semi-batch high-pressure reactor. Our approach differs from those mentioned above [19–30] in two points. In our study, the Cu/Ni weight ratio ranges from 0.02 to 0.09 whereas such a ratio was much higher in the studies reported so far (0.33 [19,20], 0.38 [21], 0.05–0.25 [23], 0.25 [24–26], 0.05–2. [27], 0.5 [28], 0.5–2.0 [29], and 0.1–9 [30]). The second difference is that the evaluation of the catalysts in the present work was performed under solvent-free conditions and the SFO volume to catalyst ratio is equal to 100 mL/1 g and the reaction time is equal to 9 h. These correspond to an LHSV value equal to 11.1 h^{-1} for experiments taken place in fixed bed reactors. These experimental conditions are very hard when compared to the corresponding ones reported in the previous works. The choice of SFO as feedstock was done by taking into account that genetically-modified sunflower grown on marginal land has been identified as sustainable biofuel source because it does not encroach upon arable lands [31].

2. Experimental

2.1. Synthesis of the Catalysts

The hydroxide precursors of the catalysts studied were prepared by co-precipitation using an aqueous solution of Al^{3+}, Ni^{2+}, and Cu^{2+} nitrate salts [Al $(NO_3)_29H_2O$, Ni $(NO_3)_26H_2O$, Cu $(NO_3)_23H_2O$, E. Merck]. This solution was in a funnel and was added drop-by-drop to a vessel containing 330 mL distilled water in which the pH was adjusted to 8 by NH_4OH. NH_4OH 30% solution (Carlo Erba Reagents) was used in a pH-control system (Metrhom) for keeping pH equal to 8 in the previously mentioned vessel during co-precipitation. Figure S2 presents the set-up used and gives more details. Even more experimental details have been reported elsewhere [6,7]. The rate of introduction of the mixed nitrate solution into the co-precipitation vessel was equal to 1.2 mL/min instead 0.7 mL/min adopted in Reference [6]. Using this rate, we prepared both the copper promoted and the non-promoted catalysts. The precipitated hydroxides were dried at 110 °C for 24 h. The dried hydroxides were decomposed to the corresponding oxides by heating them gradually under argon flow of 30 mL/min for 40 min. This time period was necessary for increasing the temperature from 25 to 400 °C. The final catalysts were then synthesized by reduction (activation) of the oxide precursors under hydrogen flow (30 mL/min) at 400 °C for 2.5 h.

2.2. Catalysts Characterization

The physicochemical properties of the catalysts were determined using various techniques. A porosimeter (Micromeritics, Tristar 3000) was used for determining the values of a specific surface area and the pore size distribution. The XRD patterns used for determining the crystal phases and the mean size of nanocrystals (Scherer's relationship) were recorded in a Brucker D8 Advance diffractometer. The catalyst morphology was determined by SEM and their composition by EDS. A SEMJEOL JSM6300 microscope with an ED spectrometer was used in all cases. The catalysts' nanostructure was investigated by TEM using a JEOL JEM-2100 system. The surface analysis of the catalysts was obtained by XPS measurements using a MAX200 (LEYBOLD/SPECS) electron spectrometer. The NH_3-TPD experiments for determining the acid sites were carried out in a laboratory-developed set up. The above techniques were applied on the final catalysts. In contrast, the H_2-TPR experiments were performed in the oxide precursors. Characterization details have been reported elsewhere [6,7,10,17,18,31–34].

2.3. Catalytic Tests

A semi-batch reactor was used in all cases. The experiments were carried out at 310 °C, H_2 pressure, a rate equal to 40 bar and 100 mL/min, respectively, and ratio of sunflower oil volume to catalyst mass equal 100 mL/1 g. The catalytic experiments were monitored for 9 h and performed under solvent-free conditions. Samples withdrawn from the reactor liquid phase were analyzed by GC (Shimadzu GC-2010, column: SUPELCO, MET-Biodiesel, l = 14 m, d = 0.53 mm, tf = 0.16 μm) and GC-MS (GC-MS-QP2010 Ultra). The accuracy of the catalytic results was determined by performing several runs twice. The results differed less than 2% in all cases. Experimental and theoretical mass balance determined for all catalytic tests differ less than ±3%. Experimental details were reported in previous contributions [6,7,10,17,18,31].

3. Results and Discussion

3.1. Catalysts Characterization

The SEM microphotographs recorded at various magnifications were similar to those published previously [6] and showed the presence of micro grains of different sizes and interparticle macro pores as well. Typical pictures for the 60NiAl and 58Ni2CuAl catalysts are presented in Figure S3. The elemental analysis, performed by EDS, indicated compositions very close to the nominal ones for the catalysts prepared. In fact, the percentage composition in nickel/copper determined for the 60NiAl, 59Ni1CuAl, 58Ni2CuAl, and 55Ni5CuAl catalysts were, respectively, equal to 59.8, 58.8/1.3, 57.3/2.3, and 55.5/6.0. A typical example of the EDS analysis concerning the 58Ni2CuAl catalyst is illustrated in Figure S4.

Figure 1 illustrates the pore volume distribution curves obtained for the catalysts studied. The curves show a mono-modal or bimodal pore size distribution in the range of 2–100 nm. The first and the second peak are centered at about 3 and 4–5 nm, respectively. More precisely, the pore size distribution curve of 60NiAl catalyst exhibits a single peak in the previously mentioned range centered at about 3 nm. The addition of a small amount of Cu (59Ni1CuAl catalyst) provoked the rise of an additional peak centered at 4–5 nm. Further increase of Cu content (58Ni2CuAl catalyst) resulted in the disappearance of the peak at 3 nm and the intensification of that centered at 4–5 nm. Adding higher Cu content (55Ni5CuAl catalyst) provoked the re-appearance of the bi-modal pore-size distribution curve but with lower intensity. In all catalysts studied, an additional broad but less intensive peak is observed in the range of pore diameter >100 nm. Thus, the solids prepared exhibit mainly mesoporous texture with the most pores concentrated in the range of 3–5 nm. This implies solids with a very high specific surface area and a very small mean pore diameter. Inspection of Table 1 shows that this is the case. The somewhat smaller value of specific surface area obtained for the 58Ni2CuAl catalyst reflects the disappearance of the pore-size distribution peak at 3.0 nm. It is notable that the addition of copper is bringing no negligible changes in the texture of the catalysts studied, which depend on the copper content.

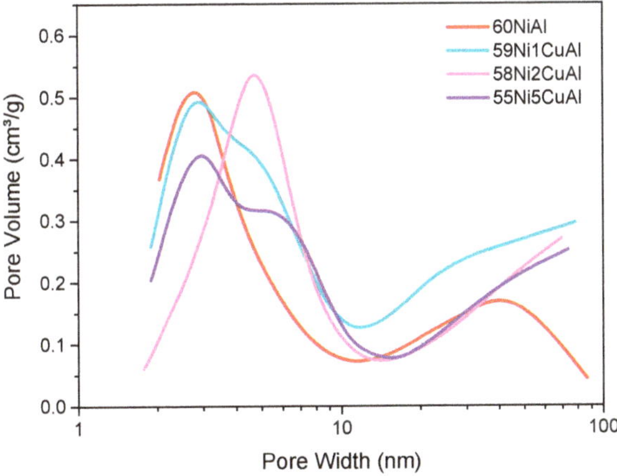

Figure 1. Pore volume distribution curves obtained for the catalysts studied.

Table 1. BET Specific Surface Area (SSA$_{BET}$), Specific Pore Volume (SPV), and Mean Pore Diameter (MPD) of the catalysts studied.

Catalyst	SSA$_{BET}$ (m^2/g)	SPV (cm^3/g)	MPD (nm)
60NiAl	247	0.30	4.84
59Ni1CuAl	285	0.43	6.03
58Ni2CuAl	192	0.33	6.91
55Ni5CuAl	230	0.33	5.74

The XRD patterns obtained after activation of the catalysts are illustrated in Figure 2. Inspection of this figure shows that NiO is predominant in the 60NiAl catalyst (main peaks 2θ: 37.2, 43.3, 62.9, and 75.4°/PDF-2 2003 # 47-1049). Moreover, the peaks assigned to metallic nickel, Ni^0, (2θ: 44.3, 51.6 and 76.1°/PDF-2 2003 # 01-1258) are also observed.

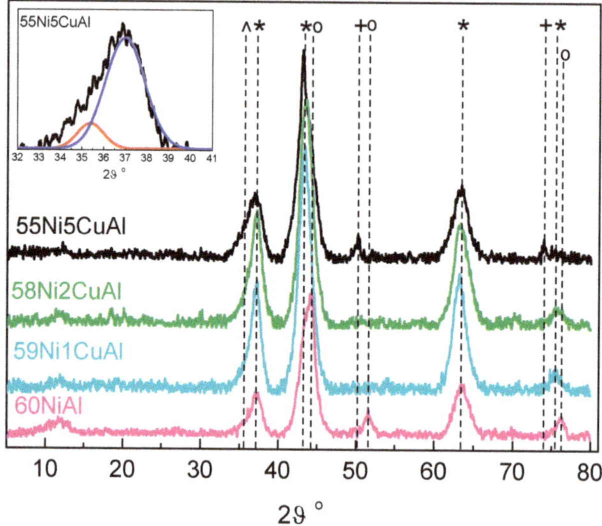

Figure 2. XRD patterns of the catalysts studied. The dashed lines correspond to the most important diffractions of the expected phases: (o) Ni^0, (*) NiO or Ni$_{0.95}$Cu$_{0.5}$O, (+) Cu^0, (^) CuO. The deconvolution of the peak at 2θ ~37° (inset) indicates the presence of CuO (red peak).

The addition of Cu to the nickel–alumina catalyst results in the disappearance of the previously mentioned peaks attributed to Ni^0. Cu favors the dispersion of Ni (as it is shown below by XPS) and the too small Ni^0 crystallites are not any more detectable by XRD. The formation of small amounts of Ni$_{0.95}$Cu$_{0.05}$O with diffraction peaks at 2θ 37.2, 43.3, 62.8, and 75.4° (PDF-2 2003 # 78-0644) cannot be excluded in the samples 59Ni1CuAl and 58Ni2CuAl while taking into account that the nominal Cu/Ni ratio in these samples is equal to 0.0157 and 0.0319, respectively, whereas this ratio in the mixed oxide is equal to 0.053. Therefore, copper is not in a sufficient quantity to form exclusively Ni$_{0.95}$Cu$_{0.5}$O. The absence of additional peaks relevant to copper phases, for instance copper/copper oxide, could be attributed to the copper entrapment in the Ni$_{0.95}$Cu$_{0.5}$O and/or to the formation of nanoparticles below the XRD identification limit. In the sample 55Ni5CuAl with nominal Cu/Ni ratio equal to 0.084, the formation of Ni$_{0.95}$Cu$_{0.5}$O is also probable, though a portion of copper is identified as metallic, Cu^0, through its characteristic peaks at 2θ 50.4 and 74.1° (PDF-2 2003 # 85-1326). The deconvolution of the asymmetric peak at 2θ ~37° indicated the presence of a monoclinic phase of CuO in this sample (see inset in Figure 2), which is revealed by its characteristic peak at 35.5° (PDF-2 2003 # 80-0076).

The absence of any peak due to alumina indicates that this oxide is largely amorphous. The nickel and copper surface speciation is further investigated by XPS.

Based on the XRD peak at 37.2°, we have calculated the mean size of the $NiO/Ni_{0.95}Cu_{0.5}O$ nanocrystals using the Scherer's relationship. It was found equal to about 5.5 nm. The TEM images of the promoted catalysts showed a rather uniform particle size distribution with a similar mean value. A typical image is illustrated in Figure S4.

Ex-situ XPS was performed on the promoted catalysts and resulted in the expected detection of Ni, Al, Cu, and O. Organic carbon from a surface contamination layer was also detected. Figure 3A shows the Ni2p3/2 spectral region of the promoted catalysts studied by XPS. The main 2p3/2 component at BE equal to 856 eV is attributed to Ni^{2+}. These spectra show that only traces of Ni^0 (less than 5% of the Ni^{2+}) are present in these catalysts, which is in good agreement with the XRD results.

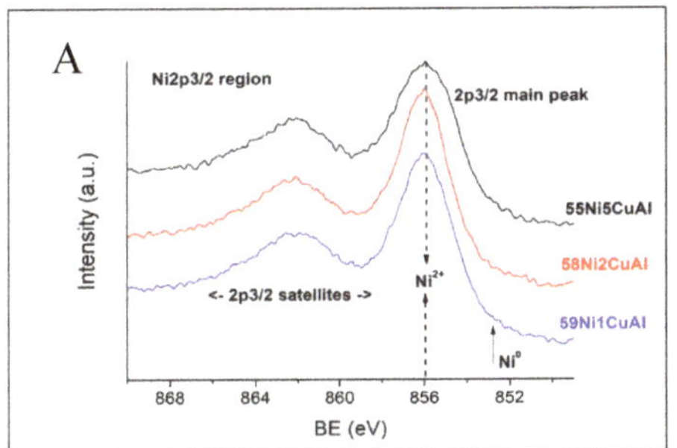

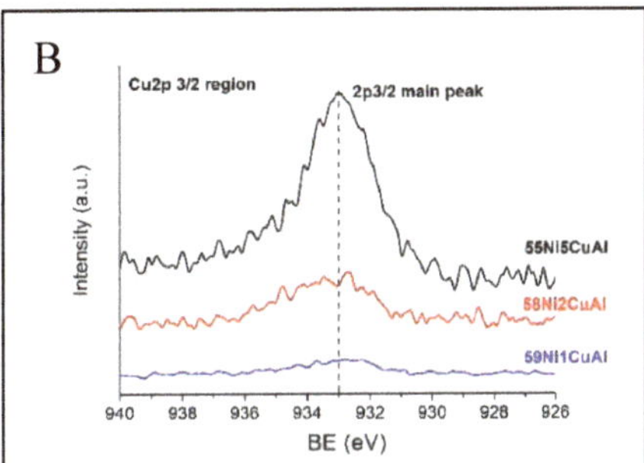

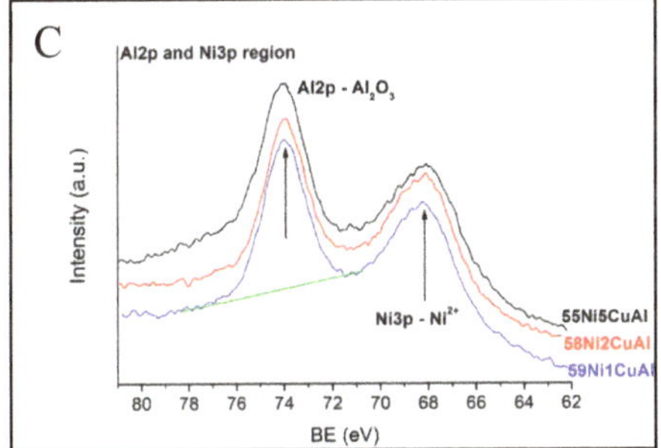

Figure 3. The XP spectra of the promoted catalysts studied: (**A**) The Ni2p3/2 spectral region, (**B**) the Cu2p3/2 spectral region, and (**C**) the Al2p–Ni3p spectral region.

Figure 3B shows the Cu2p3/2 region of the Cu-containing catalysts. Since the Cu-related spectral intensity is small, especially for the 59Ni1CuAl and 58Ni2CuAl catalysts, the Cu2p3/2 spectra for the latter two catalysts in Figure 3B represent the average of 5 and 3 scans, respectively, whereas only one scan was collected for Cu2p3/2 in the 55Ni5CuAl catalyst as well as for all other spectra presented in this section. The Cu2p3/2 peaks are relatively broad and could be described as a superposition of two states, ~932.5 eV (Cu^0 or Cu^{+1}) and ~933.7 eV (Cu^{+2}). The low BE state is predominant in the 55Ni5CuAl and 59Ni1CuAl catalysts (FWHM 2.6 and 2.8 eV, respectively), whereas the two states are comparable in the 58Ni2CuAl catalyst (FWHM 3.2 eV). The main Cu Auger LMM peaks (not shown) indicate only very small amounts of Cu^0 in the 55Ni5CuAl sample.

Figure 3C shows the broad Ni3p spectral region, which, in all cases, is superimposed by the sharp Al2p peak at a BE of 73.9 ± 0.1 eV, as expected for Al_2O_3. The area of this peak can be fairly accurately determined by drawing a linear background, as shown in Figure 3C.

The above, in conjunction with the XRD results, provide a deeper insight concerning the nickel–copper surface speciation. Ni^{2+}, Cu^0, Cu^{+1}, and Cu^{+2} phases, which is very likely among $Ni_{0.95}Cu_{0.5}O$, are present on the surface of the promoted catalysts. The expected formation of alumina was also confirmed.

Table 2 summarizes quantitative surface analysis results obtained by using XPS data (areas of Ni2p3/2, Cu2p3/2, and Al2p peak). More details were reported in Reference [6]. The results are compared with the corresponding calculated nominal atomic bulk compositions, based on the catalyst preparation procedure. The first row in Table 2 shows the corresponding data from measurements on the non-promoted 60NiAl catalyst [6]. One has to notice that the XPS-derived Ni/Al ratio is always larger than the nominal one in the bulk of the corresponding Cu-promoted catalyst (the more so, the higher the Cu content). Furthermore, this ratio is always larger than the corresponding ratio for the non-promoted catalyst (the more so, the higher the Cu content) even though the nominal Ni content in the Cu-containing catalysts should be, in all cases, somewhat smaller than that in the non-promoted one. This suggests that the presence of Cu species tends to increase the dispersion of the Ni-phase in proportion to its content, which is in agreement with the literature [23].

Table 2. Quantitative XPS results and comparison with the respective calculated nominal atomic compositions in the bulk.

Catalyst	Average Atomic Proportions (XPS) [1]	Ni/Al (Nominal)	Cu/Al (Nominal)	Cu/Ni (Nominal)	Cu/Ni (XPS) [2]
60NiAl	Al:Ni = 1:1.24	1.30	-	-	-
59Ni1CuAl	Al:Ni:Cu = 1:1.51:0.012	1.28	0.020	0.0157	0.0079
58Ni2CuAl	Al:Ni:Cu = 1:1.70:0.050	1.26	0.040	0.0319	0.0294
55Ni5CuAl	Al:Ni:Cu = 1:1.82:0.232	1.19	0.100	0.0840	0.1270

[1] Estimated uncertainty ±10% on the stated Ni and Cu values. [2] Estimated uncertainty ±20% on the stated ratios.

Another interesting observation concerns the relative surface concentration of copper in the promoted catalysts reflected in the Cu/Al ratios. The XPS Cu/Al ratio is lower than the nominal one in the 59Ni1CuAl catalyst. This presumably indicates that the much larger amount of NiO nanoparticles in this catalyst partially cover the copper species during the $Ni_{0.95}Cu_{0.5}O$ phase. This is also reflected in the value of XPS Cu/Ni atomic ratio, which is much lower than the nominal value. In the 58Ni2CuAl catalyst, the XPS Cu/Al atomic ratio is comparable to the nominal one and the value of XPS Cu/Ni atomic ratio is close to the nominal value. Both indicate that there is no considerable masking of the $Ni_{0.95}Cu_{0.5}O$ by the NiO nanoparticles. The situation is very different in the 55Ni5CuAl catalyst where both the XPS Cu/Al and Cu/Ni atomic ratios are higher than the corresponding nominal ones. These indicate that the CuO and Cu^0 nanoparticles detected by XRD may situate on the top of $Ni_{0.95}Cu_{0.5}O$ nanoparticles. Thus, the joint use of XRD and XPS characterization shed light on the relative location of the nickel and copper nanoparticles formed on the promoted samples in addition to surface speciation.

The previously mentioned ex-situ characterization indicated the formation of nickel and copper non-metallic phases. Only in the 55Ni5CuAl catalyst, metallic copper was, in addition, detected. The absence of metallic phases could be attributed either to no formation of these phases upon activation or to the extensive surface re-oxidation of the Ni^0 and Cu^0 due to the atmospheric exposure. In order to clarify this point, we performed H_2-TPR. Figure 4 shows the H_2-TPR curves of the catalysts studied.

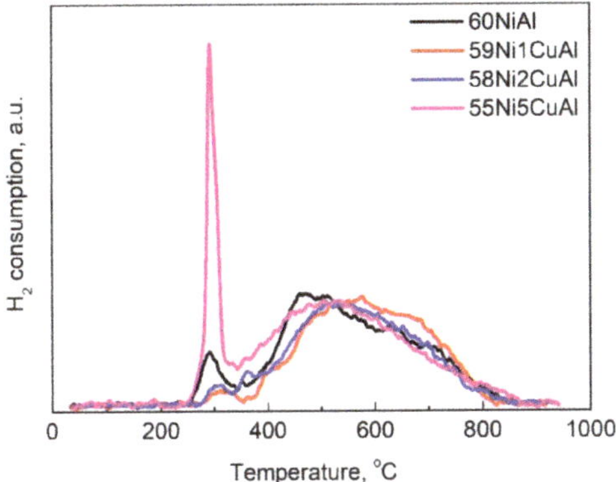

Figure 4. H_2-TPR profiles of the catalysts studied.

The H_2-TPR profile of the sample 60NiAl is characterized by two well-distinguished peaks. The first small size symmetric peak with a maximum at ~300 °C is assigned to the reduction of NiO nanoparticles weakly interacting with alumina [31,35,36]. These particles are created in the precursor state during the thermal treatment of the sample under Ar. The addition of Cu up to 2% decreases the size of this peak, which suggests a decrease of the relatively bigger NiO nanoparticles and, thus, weakly interacting with alumina. This is in line with the increase in the nickel dispersion inferred by XPS. The very high peak at about 300 °C in the sample with the higher Cu loading (55Ni5CuAl) is attributed to the reduction of CuO to Cu^0 [23,27–29]. There is no doubt that both the previously mentioned NiO and CuO reductions occur upon activation taking place at 400 °C. It is plausible assuming that the metallic nickel (copper) formed is entirely (partially) re-oxidized upon atmosphere exposure, which is in agreement with the XRD and XPS results. The second broad peak is assigned to various Ni-oxo species strongly interacting with the alumina and resembling the hardly reduced $NiAl_2O_4$ phase [31,35,36]. The addition of Cu is causing a slight shift of the maximum of the broad peak toward higher temperatures, which likely indicates the increase in the $NiO/Ni_{0.95}Cu_{0.5}O$ dispersion deduced by XPS. Taking into account that (i) the low temperature extreme of the broad reduction band is located below 400 °C, (ii) the catalysts activation-reduction time (2.5 h), and (iii) the transient character of the H_2-TPR method, we are arguing that the reduction of the $NiO/Ni_{0.95}Cu_{0.5}O$ and CuO species present in the precursor catalysts are reduced in great extent upon activation. Then these are re-oxidized upon atmospheric exposure. This is in line with the literature. In fact, Miao et al. [30] reported that the oxidic precursors are reduced at 420 °C for 2 h into the corresponding metallic catalysts, comprised from Ni^0, Cu^0, and NiCu alloy-supported species. In conclusion, the NiO, CuO, and $Ni_{0.95}Cu_{0.5}O$ species deduced by the Ex-situ characterization had been presumably transformed in a great extent into Ni^0, Cu^0, and a Ni-Cu alloy very rich in nickel catalysts upon activation. To further investigate this point, we are performing the following experiment for the 58Ni2CuAl catalyst, taken as an example. The precursor catalyst was activated in the TPR set–up at 400 °C for 2.5 h, the temperature was reduced at 100 °C under hydrogen flow, and the H_2-TPR experiment was performed in situ for the activated catalyst (Figure 5). A comparison of the H_2-TPR profiles of the precursor and the activated catalyst clearly shows that inception point of catalyst reduction was shifted from ~200 to ~450 °C after catalyst activation whereas the amount of the reduced species was considerably reduced. This corroborates our previous assumption that the NiO, CuO, and $Ni_{0.95}Cu_{0.5}O$ species deduced by the Ex-situ characterization had been transformed in a great extent into Ni^0, Cu^0, and a Ni-Cu alloy very rich in nickel upon activation. This is not the case for the like $NiAl_2O_4$ surface phase since the reduction above 700 °C is similar for the precursor and activated catalyst. This phase seems to be well dispersed and, thus, it is not detectable by XRD. In this point, one may wonder whether the Ni^0, Cu^0, and a Ni-Cu species re-oxidized upon exposure in air are actually present in the functional catalysts working under a very reducible atmosphere (hydrogen pressure 40 bar). There is strong evidence that it is actually the case.

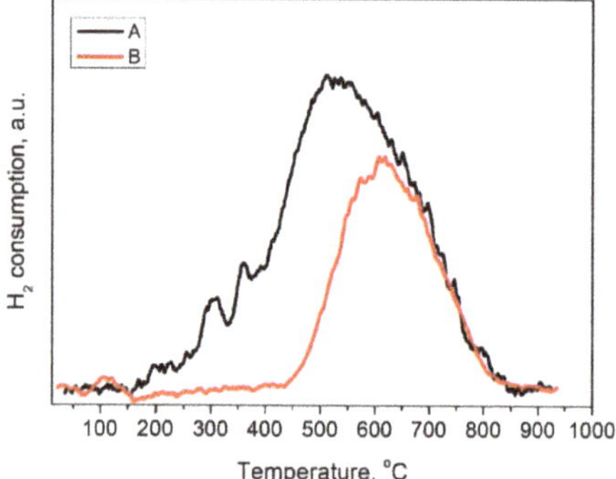

Figure 5. H_2-TPR profiles of the 58Ni2CuAl precursor catalyst (**A**) and of the same catalyst after its in-situ activation at 400 °C for 2.5 h (**B**).

In fact, Totong et al. [37] have recorded the H_2-TPR profiles after chemical activation (reduction by $NaBH_2$) of nickel catalyst (10 wt. % Ni) supported on CeO_2-ZrO_2 and its exposure in air. The H_2-TPR profile exhibited re-reduction peaks. Moreover, the formation of the Ni-Cu alloy has been reported several times in the literature [19–23,28–30]. The formation of the Cu–Ni alloy favors SDO and inhibits cracking of the C–C bonds, which leads to enhanced catalytic performance [23,28,29]. Extending the picture, we had outlined the local arrangement of nickel and copper oxide species into the reduced ones. We are visualizing the surface speciation in the functional catalysts. In the 59Ni1CuAl catalyst, the much larger amount of Ni^0 nanoparticles covers partially the Ni-Cu alloy, whereas, in the 58Ni2CuAl catalyst, there is no considerable masking of this Ni-Cu alloy by the Ni^0 nanoparticles. Lastly, in the 58Ni5CuAl catalyst, a portion of Cu^0 nanoparticles partially cover the Ni-Cu alloy ones.

The acidity of the catalysts has been studied by NH_3-TPD. Figure 6 reveals that all catalysts exhibit mainly weak (desorption temperature range <300 °C) and relatively low moderate (desorption temperature range 300–450 °C) and very low strong (desorption temperature range >450 °C) acidity [17,38,39]. A small addition of Cu (59Ni1CuAl) increases total acidity. However, this trend is reversed as the Cu loading is further increased. It is interesting that the sample with the maximum copper content exhibits considerably low weak/moderate acidity and relatively high strong acidity.

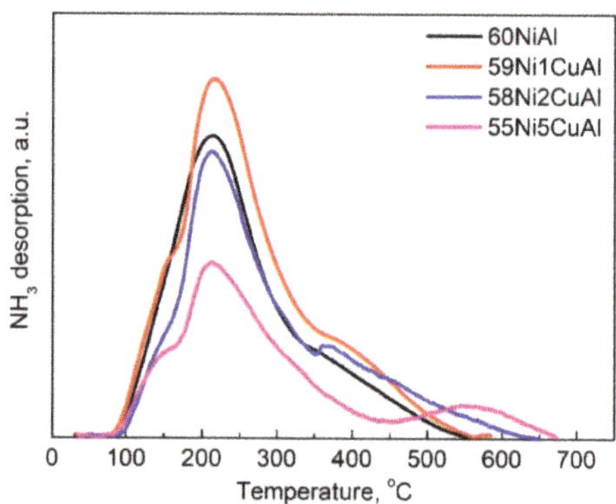

Figure 6. NH_3-TPD curves of the catalysts studied.

3.2. Catalysts Evaluation

A representative chromatogram taken after sampling the liquid phase of the reactor is depicted in Figure 7.

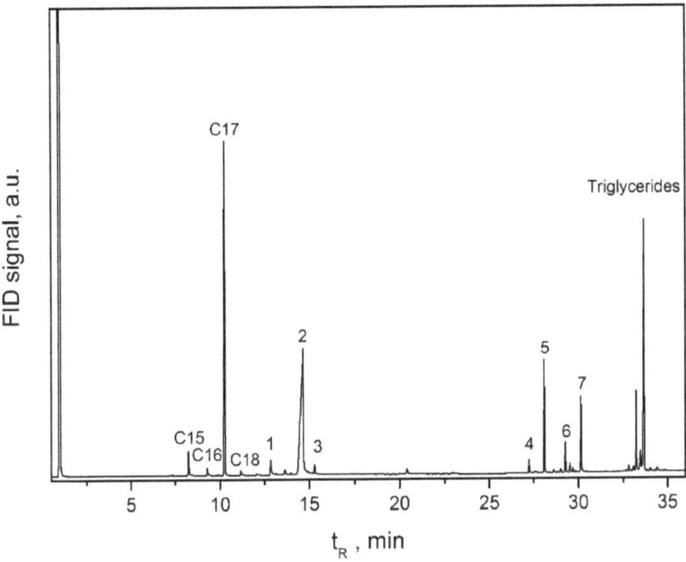

Figure 7. A typical gas chromatogram of a liquid sample received from the reactor upon transformation of SFO into green over the 58Ni2CuAl catalyst (reaction time = 4 h): C15. Pentadecane, C16. Hexadecane, C17. Heptadecane, C18. Octadecane, 1. Palmitic acid, 2. Stearic acid/ethyl stearate, 3. Propyl stearate, 4. Palmityl stearate, 5. Stearyl stearate, 6. 2(octadecyloxy) ethyl stearate, 7. Distearine. (Reaction conditions: 310 °C, 40 bar hydrogen pressure, hydrogen flow rate = 100 mL/min (STP), reactant volume to catalyst mass ratio = 100 mL/1 g).

Inspection of the chromatograms in conjunction with the study of GC-MS spectra showed the presence of the non–reacted SFO and normal alkanes in the diesel range (n-C17, n-C18, n-C15, and n-C16). Hydrocarbons with a smaller number of carbon atoms are not detected in the liquid phase of the reaction mixture. We have detected intermediate compounds such as acids (palmitic and stearic acid), propyl and ethyl esters (ethyl stearate and propyl stearate), bigger esters (palmityl stearate, stearyl stearate, 2(octadecyloxy) ethyl stearate), and saturated diglycerides (distearine). In the gas phase, we have detected propane, CO, CO_2, and CH_4. The same products were detected in the presence of the non-promoted and the promoted catalysts, which suggests no considerable influence of copper on the entire SDO reaction network described in the introduction.

The variation with time of the conversion of SFO and the composition of the liquid reaction mixture in hydrocarbons, acids, and esters, determined over the most active catalyst 58Ni2CuAl, is illustrated in Figure 8.

Inspection of this figure shows that the composition in total hydrocarbons and each hydrocarbon separately increase monotonically with time. In contrast, the composition in the intermediate total acids and esters increases initially with time and then it decreases. Similar kinetic curves were obtained for all the catalysts evaluated in the present work. This suggests no considerable influence of copper on the SDO mechanistic scheme described in the introduction for the nickel-based catalysts.

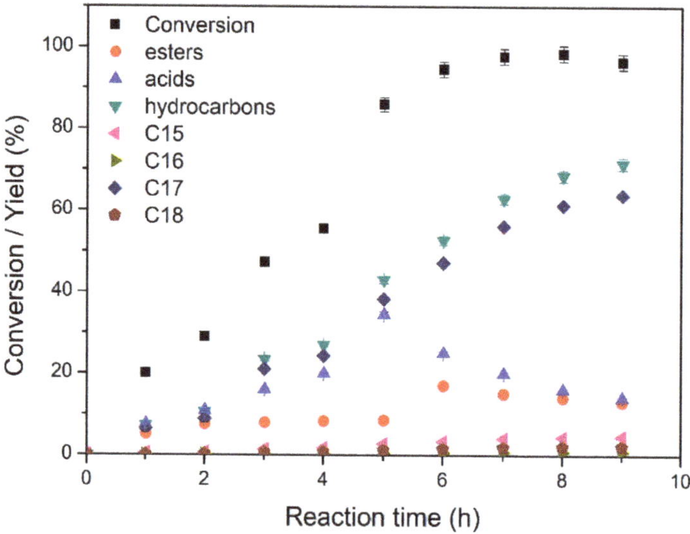

Figure 8. Kinetics of the SFO selective deoxygenation over the most active catalyst (58Ni2CuAl). (Reaction conditions: 310 °C, 40 bar hydrogen pressure, hydrogen flow rate = 100 mL/min (STP), reactant volume to catalyst mass ratio = 100 mL/1 g).

Figure 9 presents the conversion of SFO, and the composition of the liquid reaction mixture in total n-alkanes, acids, and esters, determined after 9 h of the reaction.

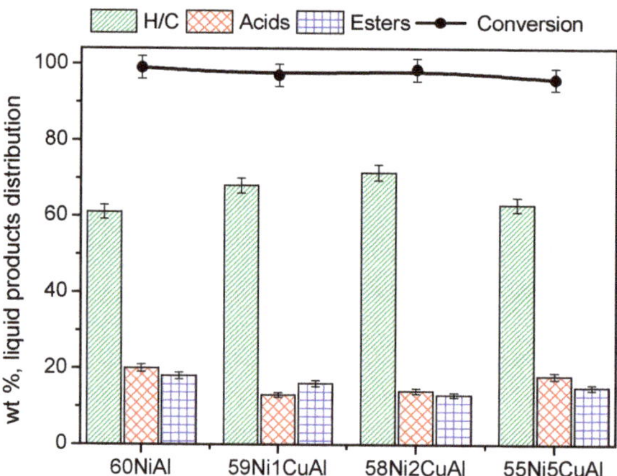

Figure 9. Evaluation parameters obtained for the SDO of SFO over the catalysts studied at reaction time equal to 9 h (Reaction conditions: 310 °C, 40 bar hydrogen pressure, hydrogen flow rate = 100 mL/min (STP), reactant volume to catalyst mass ratio = 100 mL/1 g).

A very high conversion is obtained in all cases. The most important intermediate products are esters and fatty acids with their compositions in the reaction mixture in the range of 13–18% and 13–20%, respectively. The most important finding is that copper considerably increases the composition of the liquid reaction mixture in total hydrocarbons in the diesel range, which is the most significant evaluation parameter from the practical point-of-view. The effect is more pronounced in the presence of 58Ni2CuAl catalyst, where an increase of 17.2% has been obtained with respect to the non-promoted

60NiAl catalyst. Therefore, the promoting action of copper is clearly emerged for the first time for catalysts with a very small Cu/Ni weight ratio (0.02–0.09) operating under solvent-free conditions and very high SFO volume-to-catalyst ratio (100 mL/1 g) and a reaction time of 9 h. These correspond to an LHSV value equal to 11.1 h^{-1} for experiments taken place in fixed bed reactors. Therefore, the values of the composition of the liquid reaction mixture in total hydrocarbons obtained in the present experiments correspond to the mean value obtained for experiments taken place in a fixed bed reactor at LHSV value equal to 1 h^{-1} and time on stream of 100 h. This is in conjunction with the almost linear increase of this evaluation parameter (Figure 8) with time providing a strong evidence for the relatively high stability of the catalysts tested.

At this point, it should be noted that the commercially produced green diesel initially contains saturated hydrocarbons with carbon atoms in the range of C15–C18, mainly C17 and C18, as in the present work. These linear hydrocarbons improve the ignition properties (cetane number) of fuel, but reduce its flow characteristics. Thus, an effort has been undertaken to improve the latter. Yeletsky et al. [40] have reviewed this issue, which presented recent advances. One may also notice that the yield of total hydrocarbons obtained over the most active catalyst studied was 71.5% (the rest being mainly high molecular weight esters and free fatty acids (Figure 9)). However, the hydrocarbons yield can be very easily increased to 100% by a slight decrease in the "volume of SFO to catalyst mass" ratio.

Figure 10 illustrates the composition of the liquid product in n-alkanes obtained over the catalysts studied at 9 h of the reaction.

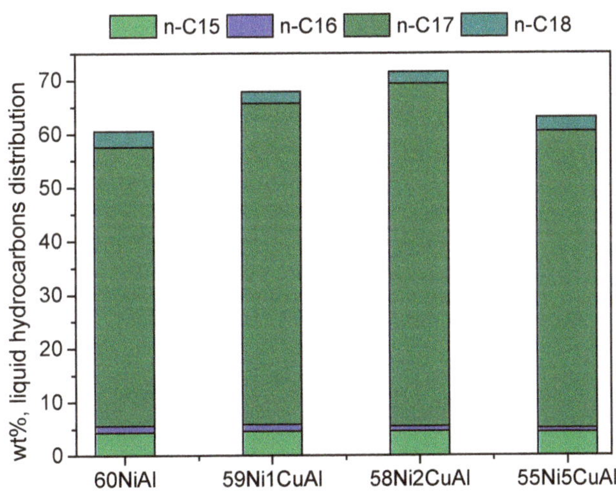

Figure 10. Composition of the liquid product in n-alkanes with odd and even numbers of carbon atoms obtained over the catalysts studied at reaction time equal to 9 h (Reaction conditions: 310 °C, 40 bar hydrogen pressure, hydrogen flow rate = 100 mL/min (STP), reactant volume to catalyst mass ratio = 100 mL/1 g).

It is seen that the content of the reaction mixture in hydrocarbons with an odd number of carbon atoms (n-C15 and mainly n-C17) are considerably higher than that of the corresponding ones with even carbon atoms (n-C16 and n-C18). In fact, the values obtained for the ratios (n-C17/(n-C17+n-C18)) and (n-C15/(n-C15+n-C16)) are, respectively, centered in the range of 0.97–0.95 and 0.86–0.77, which indicates that SDO mainly proceeds via deH$_2$O-deCO rather than through deH$_2$O as expected for nickel catalysts [5–8,10,17,18,31]. The above indicate that the presence of small amounts of copper does not disturb the SDO mechanism.

It seems to us reasonable to relate the copper-promoting action to the increase of nickel dispersion brought about by copper, which is deduced by XPS. However, the monotonous increase of Ni dispersion with the Cu content (Table 2) is not compatible with the volcano-like trend of catalytic performance and its maximization over the catalyst with the medium Cu content (58Ni2CuAl). The inferred formation of a Ni-Cu alloy on the base of catalyst characterization should be another crucial factor [19–23,28–30]. The close location of Ni and Cu atoms inside the alloy allows the development of electronic interactions between these two elements, which may influence the intensity of adsorption of hydrogen and several key intermediates and, thus, the catalytic performance to hydrocarbons [30]. It is known that the intensity of adsorption for hydrogen, oxygen, carbon, and several organic molecules is lower on copper when compared to nickel [41] and this could explain the different catalytic behavior exhibited from monometallic nickel and Ni-Cu alloys in several catalytic reactions [42]. As already concluded on the surface of the 59Ni1CuAl and 55Ni5CuAl catalysts, a portion of the more active Ni-Cu alloy nanoparticles is covered by Ni^0 and Cu^0 nanoparticles, respectively. On the contrary, there is no considerable masking of this Ni-Cu alloy in the 58Ni2CuAl catalyst explaining its highest catalytic performance. The latter could be secondarily attributed to the mono-modal pore size distribution exhibited by this catalyst, centered at about 4.5 nm (Figure 1), which presumably facilitates the reactants/products' mass transfer. The relatively low performance of the 55Ni5CuAl catalyst with respect to the 59Ni1CuAl and 58Ni2CuAl catalysts could be partly attributed to its remarkably low weak/moderate acidity and the relatively high strong acidity. The former favors SDO whereas the second catalytic cracking results in coking [5].

A direct comparison of the present catalytic results with those of the literature ones is not an easy task. This is due to the high diversity of the set ups (fixed bed, batch, semi-batch), experimental conditions (temperatures: 250–400 °C, hydrogen pressures: 5–40 bar, equivalent LHSV values: 1–7 h^{-1}), and feedstocks (biodiesel, methyl palmitate, ethyl caprate, tristearin, stearic acid, yellow grease, hemp seed oil, waste free fatty acids, brown grease, oleic acid, methyl laurate, and sunflower oil) used in the relevant works [19–30]. However, some useful conclusions could be drawn by comparing the present results with the literature results. First, it should be mentioned that, even though the non-promoted catalyst (60NiAl) is very active when compared to the current catalysts, a yield of 61.2% of total hydrocarbons was obtained over this catalyst working under solvent-free conditions and an equivalent LHSV value equal to 11.1 h^{-1}. In spite of this high activity, the presence of a very small amount of copper is sufficient to considerably increase the previously mentioned yield. The maximum promoting action of copper in the present work is obtained in the 58Ni2CuAl catalyst with a Cu/Ni weight ratio equal to 0.034, which is much smaller to those obtained in the other relevant works [19–30]. We are attributing this difference to the preparation method followed in the present study (controlled co-precipitation). This method results initially in the formation of layered double hydroxide precipitates in which the Ni^{2+}, Cu^{2+}, and Al^{3+} ions are atomically mixed in the basal layers of the double hydroxide [43–45]. This, in turn, facilitates the well mixing of very small alumina, NiO, CuO, and $Ni_{0.95}Cu_{0.5}O$ particles formed in the subsequent thermal step (heating under argon) and, thus, the well mixing and high dispersion of the Ni^0, Cu^0, and Ni-Cu alloy rich in nickel is formed upon activation. The high-level synergy between the Ni^0 hydrogenation sites and Cu^0 sites inside the Ni-Cu alloy seems to be the main reason for the promoting action of copper.

As already mentioned, the presence of copper does not change practically in the SDO network followed in metallic nickel catalysts where the deH$_2$O-deCO pathway predominates. This is in agreement with the literature. In fact, copper has a weak effect, if any, on SDO mechanism. The Ni-Cu bimetallic catalysts still prefer catalyzing the deH$_2$O-deCO pathway [19,23]. A shift of the mechanism from the deH$_2$O-deCO to the deH$_2$O pathway takes place in copper-promoted catalysts above a critical Cu/Ni weight ratio much higher than those studied in the present work. For example, Miao et al. [30] have been recently reported where the SDO network changes from deH$_2$O-deCO to the deH$_2$O pathway at a Cu/Ni weight ratio higher than 2.3. The relatively high concentration of the intermediate esters determined in the present study (Figure 9) was compared to most of the relevant works [19–30], which can be effortlessly related to the solvent-free conditions adopted in the present study. This facilitates the bimolecular esterification between the intermediates' fatty acids and fatty alcohols.

4. Conclusions

Copper doping considerably increases the catalytic performance of the nickel–alumina co-precipitated catalysts (60 wt. % Ni) for the transformation of SFO into green diesel whereas it does not affect the SDO reaction network. The effect is more pronounced in the catalyst containing 2 wt. % copper (58Ni2CuAl), where a 17.2% increase of green diesel content in the liquid product has been achieved with respect to the non-promoted catalyst, 60NiAl. The promoting action of copper is demonstrated for the first time for catalysts with a very small Cu/Ni weight ratio (0.02–0.09). This was attributed to the increase in the nickel dispersion caused by copper and to the formation of a Ni-Cu alloy very rich in nickel, which is considered more active than nickel. In the 59Ni1CuAl and 55Ni5CuAl catalysts, a portion of the Ni-Cu alloy is covered by Ni^0 and Cu^0 nanoparticles, respectively. On the contrary, there is no considerable masking of this Ni-Cu alloy in the 58Ni2CuAl catalyst, which explains its highest catalytic performance. The relatively low performance of the 55Ni5CuAl catalyst could be partly attributed to its remarkably low weak/moderate acidity and the relatively high strong acidity. The former favors SDO whereas the second catalytic cracking results in coking.

Supplementary Materials: The following are available online at http://www.mdpi.com/1996-1073/13/14/3707/s1. Figure S1: SDO reaction network. Figure S2: Presentation of the setup used for co-precipitation. Figure S3: SEM images. Figure S4: EDS spectrum. Figure S5: TEM image.

Author Contributions: Conceptualization, K.B., C.K. and A.L.; Methodology, M.G. and E.K.; Software, S.L.; Validation, E.K., S.L. and E.S.; Investigation, M.G., E.K. and E.S.; Data Curation, M.G., E.K. and E.S.; Writing-Original Draft Preparation, A.L.; Writing-Review & Editing, K.B.; Visualization, K.B., C.K. and A.L.; Supervision, K.B. and S.L. All authors have read and agreed to the published version of the manuscript.

Funding: This research received no external funding.

Conflicts of Interest: The authors declare no conflict of interest.

References

1. Lycourghiotis, A.; Kordulis, C.; Lycourghiotis, S. *Beyond Fossil Fuels: The Return Journey to Renewable Energy;* Crete University press: Herakleion, Greece, 2017.

2. Armaroli, N.; Balzani, V. The Future of Energy Supply: Challenges and Opportunities. *Angew. Chem. Int. Ed.* **2007**, *46*, 52–66. [CrossRef] [PubMed]

3. Thomas, J.M.; Harris, K.D.M. Some of tomorrow's catalysts for processing renewable and non-renewable feedstocks, diminishing anthropogenic carbon dioxide and increasing the production of energy. *Energy Environ. Sci.* **2016**, *9*, 687–708. [CrossRef]

4. Frusteri, F.; Aranda, D.; Bonura, G. (Eds.) *Sustainable Catalysis for Biorefineries;* Green Chemistry Series No 56; RSC: Croydon, UK, 2018. [CrossRef]

5. Kordulis, C.; Bourikas, K.; Gousi, M.; Kordouli, E.; Lycourghiotis, A. Development of nickel based catalysts for the transformation of natural triglycerides and related compounds into green diesel: A critical review. *Appl. Catal. B Environ.* **2016**, *181*, 156–196. [CrossRef]

6. Gousi, M.; Andriopoulou, C.H.; Bourikas, K.; Ladas, S.; Sotiriou, M.; Kordulis, C.; Lycourghiotis, A. Green diesel production over nickel-alumina co-precipitated catalysts. *Appl. Catal. A Gen.* **2017**, *536*, 45–56. [CrossRef]

7. Zafeiropoulos, G.; Nikolopoulos, N.; Kordouli, E.; Sygellou, L.; Bourikas, K.; Kordulis, C.; Lycourghiotis, A. Developing Nickel–Zirconia Co-Precipitated Catalysts for Production of Green Diesel. *Catalysts* **2019**, *9*, 210. [CrossRef]

8. Lycourghiotis, S.; Kordouli, E.; Sygellou, L.; Bourikas, K.; Kordulis, C. Nickel catalysts supported on palygorskite for transformation of waste cooking oils into green diesel. *Appl. Catal. B Environ.* **2019**, *259*, 118059. [CrossRef]

9.   Hongloi, N.; Prapainainar, P.; Seubsai, A.; Sudsakorn, K.; Prapainainar, C. Nickel catalyst with different supports for green diesel production. *Energy* **2019**, *182*, 306–320. [CrossRef]

10. Nikolopoulos, I.; Kogkos, G.; Kordouli, E.; Bourikas, K.; Kordulis, C.; Lycourghiotis, A. Waste cooking oil transformation into third generation green diesel catalyzed by nickel - Alumina catalysts. *Mol. Catal.* **2020**, *482*, 110697. [CrossRef]

11. Ochoa-Hernández, C.; Coronado, J.M.; Serrano, D.P. Hydrotreating of Methyl Esters to Produce Green Diesel over Co- and Ni-Containing Zr-SBA-15 Catalysts. *Catalysts* **2020**, *10*, 186. [CrossRef]

12. Papanikolaou, G.; Lanzafame, P.; Giorgianni, G.; Abate, S.; Perathoner, S.; Centi, G. Highly selective bifunctional Ni zeo-type catalysts for hydroprocessing of methyl palmitate to green diesel. *Catal. Today* **2020**, *345*, 14–21. [CrossRef]

13. Ameen, M.; Azizan, M.T.; Ramli, A.; Yusup, S.; Abdullah, B. The effect of metal loading over Ni/γ-Al2O3 and Mo/γ-Al2O3 catalysts on reaction routes of hydrodeoxygenation of rubber seed oil for green diesel production. *Catal. Today* **2019**, in press. [CrossRef]

14. Srifa, A.; Kaewmeesri, R.; Fang, C.; Itthibenchapong, V.; Faungnawakij, K. $NiAl_2O_4$ spinel-type catalysts for deoxygenation of palm oil to green diesel. *Chem. Engin. J.* **2018**, *345*, 107–113. [CrossRef]

15. Feng, F.; Shang, Z.; Wang, L.; Zhang, X.; Liang, X.; Wang, Q. Structure-sensitive hydro-conversion of oleic acid to aviation-fuel-range-alkanes over alumina-supported nickel catalyst. *Catal. Commun.* **2020**, *134*, 105842. [CrossRef]

16. Afshar Taromi, A.; Kaliaguine, S. Hydrodeoxygenation of triglycerides over reduced mesostructured Ni/γ-alumina catalysts prepared via one-pot sol-gel route for green diesel production. *Appl. Catal. A Gen.* **2018**, *558*, 140–149. [CrossRef]

17. Kordouli, E.; Pawelec, B.; Bourikas, K.; Kordulis, C.; Fierro, J.L.G.; Lycourghiotis, A. Mo promoted Ni-Al_2O_3 co-precipitated catalysts for green diesel production. *Appl. Catal. B Environ.* **2018**, *229*, 139–154. [CrossRef]

18. Gousi, M.; Kordouli, E.; Bourikas, K.; Simianakis, E.; Ladas, S.; Panagiotou, G.D.; Kordulis, C.; Lycourghiotis, A. Green diesel production over nickel-alumina nanostructured catalysts promoted by zinc. *Catal. Today* **2019**, in press. [CrossRef]

19. Yakovlev, V.A.; Khromova, S.A.; Sherstyuk, O.V.; Dundich, V.O.; Ermakov, D.Y.; Novopashina, V.M.; Lebedev, M.Y.; Bulavchenko, O.; Parmon, V.N. Development of new catalytic systems for upgraded bio-fuels production from bio-crude-oil and biodiesel. *Catal. Today* **2009**, *144*, 362–366. [CrossRef]

20. Dundich, V.O.; Khromova, S.A.; Ermakov, D.Y.; Lebedev, M.Y.; Novopashina, V.M.; Sister, V.G.; Yakimchuk, A.I.; Yakovlev, V.A. Nickel Catalysts for the Hydrodeoxygenation of Biodiesel. *Kinet. Catal.* **2010**, *51*, 704–709. [CrossRef]

21. Kukushkin, R.G.; Bulavchenko, O.A.; Kaichev, V.V.; Yakovlev, V.A. Influence of Mo on catalytic activity of Ni-based catalysts in hydrodeoxygenation of esters. *Appl. Catal. B Environ.* **2015**, *163*, 531–538. [CrossRef]

22. Kukushkin, R.G.; Eletskii, P.M.; Bulavchenko, O.A.; Saraev, A.A.; Yakovlev, V.A. Studying the Effect of Promotion with Copper on the Activity of the Ni/Al_2O_3 Catalyst in the Process of Ester Hydrotreatment. *Catal. Indust.* **2019**, *11*, 198–207. [CrossRef]

23. Loe, R.; Santillan-Jimenez, E.; Morgan, T.; Sewell, L.; Ji, Y.; Jones, S.; Isaacs, M.A.; Lee, A.F.; Crocker, M. Effect of Cu and Sn promotion on the catalytic deoxygenation of model and algal lipids to fuel-like hydrocarbons over supported Ni catalysts. *Appl. Catal. B Environ.* **2016**, *191*, 147–156. [CrossRef]

24. Santillan-Jimenez, E.; Loe, R.; Garrett, M.; Morgan, T.; Crocker, M. Effect of Cu promotion on cracking and methanation during the Ni-catalyzed deoxygenation of waste lipids and hemp seed oil to fuel-like hydrocarbons. *Catal. Today* **2018**, *302*, 261–271. [CrossRef]

25. Loe, R.; Lavoignat, Y.; Maier, M.; Abdallah, M.; Morgan, T.; Qian, D.; Pace, R.; Santillan-Jimenez, E.; Crocker, M. Continuous Catalytic Deoxygenation of Waste Free Fatty Acid-Based Feeds to Fuel-Like Hydrocarbons Over a Supported Ni-Cu Catalyst. *Catalysts* **2019**, *9*, 123. [CrossRef]

26. Silva, G.C.R.; Qian, D.; Pace, R.; Heintz, O.; Caboche, G.; Santillan-Jimenez, E.; Crocker, M. Promotional Effect of Cu, Fe and Pt on the Performance of Ni/Al_2O_3 in the Deoxygenation of Used Cooking Oil to Fuel-Like Hydrocarbons. *Catalysts* **2020**, *10*, 91. [CrossRef]

27. Jing, Z.; Zhang, T.; Shang, J.; Zhai, M.; Yang, H.; Qiao, C.; Ma, X. Influence of Cu and Mo components of γ-Al_2O_3 supported nickel catalysts on hydrodeoxygenation of fatty acid methyl esters to fuel-like hydrocarbons. *J. Fuel Chem. Technol.* **2018**, *46*, 427–440. [CrossRef]

28. Zhang, Z.; Chen, H.; Wang, C.; Chen, K.; Lu, X.; Ouyang, P.; Fu, J. Efficient and stable Cu-Ni/ZrO$_2$ catalysts for in situ hydrogenation and deoxygenation of oleic acid into heptadecane using methanol as a hydrogen donor. *Fuel* **2018**, *230*, 211–217. [CrossRef]

29. Zhang, Z.; Yang, Q.; Chen, H.; Chen, K.; Lu, X.; Ouyang, P.; Fu, J.; Chen, J.G. In situ hydrogenation and decarboxylation of oleic acid into heptadecane over a Cu–Ni alloy catalyst using methanol as a hydrogen carrier. *Green Chem.* **2018**, *20*, 197–205. [CrossRef]

30. Miao, C.; Zhou, G.; Chen, S.; Xie, H.; Zhang, X. Synergistic effects between Cu and Ni species in NiCu/γ-Al$_2$O$_3$ catalysts for hydrodeoxygenation of methyl laurate. *Renew. Energy* **2020**, *153*, 1439–1454. [CrossRef]

31. Kordouli, E.; Sygellou, L.; Kordulis, C.; Bourikas, K.; Lycourghiotis, A. Probing the synergistic ratio of the NiMo/γ-Al$_2$O$_3$ reduced catalysts for the transformation of natural triglycerides into green diesel. *Appl. Catal. B Environ.* **2017**, *209*, 12–22. [CrossRef]

32. Makarouni, D.; Lycourghiotis, S.; Kordouli, E.; Bourikas, K.; Kordulis, C.; Dourtoglou, V. Transformation of limonene into p-cymene over acid activated natural mordenite utilizing atmospheric oxygen as a green oxidant: A novel mechanism. *Appl. Catal. B Environ.* **2018**, *224*, 740–750. [CrossRef]

33. Lycourghiotis, S.; Makarouni, D.; Kordouli, E.; Bourikas, K.; Kordulis, C.; Dourtoglou, V. Activation of natural mordenite by various acids: Characterization and evaluation in the transformation of limonene into p-cymene. *Mol. Catal.* **2018**, *450*, 95–103. [CrossRef]

34. Lycourghiotis, S.; Makarouni, D.; Kordouli, E.; Bourikas, K.; Kordulis, C.; Dourtoglou, V. Transformation of limonene into high added value products over acid activated natural montmorillonite. *Catal. Today* **2019**, in press. [CrossRef]

35. Liu, Q.; Gao, J.; Gu, F.; Lu, X.; Liu, Y.; Li, H.; Zhong, Z.; Liu, B.; Xu, G.; Su, F. One-pot synthesis of ordered mesoporous Ni–V–Al catalysts for CO methanation. *J. Catal.* **2015**, *326*, 127–138. [CrossRef]

36. Tribalis, A.; Panagiotou, G.D.; Bourikas, K.; Sygellou, L.; Kennou, S.; Ladas, S.; Lycourghiotis, A.; Kordulis, C. Ni Catalysts Supported on Modified Alumina for Diesel Steam Reforming. *Catalysts* **2016**, *6*, 11. [CrossRef]

37. Totong, S.; Daorattanachai, P.; Laosiripojana, N.; Idem, R. Catalytic depolymerization of alkaline lignin to value-added phenolic-based compounds over Ni/CeO$_2$-ZrO$_2$ catalyst synthesized with a one-step chemical reduction of Ni species using NaBH4 as the reducing agent. *Fuel Process. Technol.* **2020**, *198*, 106248. [CrossRef]

38. Zhang, D.; Liu, W.-Q.; Liu, Y.-A.; Etim, U.J.; Liu, X.-M.; Yan, Z.-F. Pore confinement effect of MoO3/Al2O3 catalyst for deep hydrodesulfurization. *Chem. Eng. J.* **2017**, *330*, 706–717. [CrossRef]

39. Wagenhofer, M.F.; Baráth, E.; Gutiérrez, O.Y.; Lercher, J.A. Carbon–carbon bond scission pathways in the deoxygenation of fatty acids on transition-metal sulfides. *ACS Catal.* **2017**, *7*, 1068–1076. [CrossRef]

40. Yeletsky, P.M.; Kukushkin, R.G.; Yakovlev, V.A.; Chen, B.H. Recent advances in one-stage conversion of lipid-based biomass-derived oils into fuel components—Aromatics and isomerized alkanes. *Fuel* **2020**, *278*, Art.N. 118255. [CrossRef]

41. Wu, Q.; Duchstein, L.D.L.; Chiarello, G.L.; Christensen, J.M.; Damsgaard, C.H.D.; Elkjar, C.H.F.; Wagner, J.B.; Temel, B.; Grunwaldt, J.-D.; Jensen, A.D. In Situ Observation of Cu–Ni Alloy Nanoparticle Formation by X-Ray Diffraction, X-Ray Absorption Spectroscopy, and Transmission Electron Microscopy: Influence of Cu/Ni Ratio. *Chemcatchem* **2014**, *6*, 301–310. [CrossRef]

42. Saw, E.T.; Oemar, U.; Tan, X.R.; Du, Y.; Borgna, A.; Hidajat, K.; Kawi, S. Bimetallic Ni–Cu catalyst supported on CeO2 for high-temperature water–gas shift reaction: Methane suppression via enhanced CO adsorption. *J. Catal.* **2014**, *314*, 32–46. [CrossRef]

43. Fan, G.; Li, F.; Evans, D.G.; Duan, X. Catalytic applications of layered double hydroxides: Recent advances and perspectives. *Chem. Soc. Rev.* **2014**, *43*, 7040–7066. [CrossRef] [PubMed]

44. Li, C.; Wei, M.; Evans, D.G.; Duan, X. Layered double hydroxide-based nanomaterials as highly efficient catalysts and adsorbents. *Small* **2014**, *10*, 4469–4486. [CrossRef] [PubMed]

45. He, S.; An, Z.; Wei, M.; Evans, D.G.; Duan, X. Layered double hydroxide-based catalysts: Nanostructure design and catalytic performance. *Chem. Commun.* **2013**, *49*, 5912–5920. [CrossRef] [PubMed]

MDPI

St. Alban-Anlage 66

4052 Basel

Switzerland

Tel. +41 61 683 77 34

Fax +41 61 302 89 18

www.mdpi.com

Energies Editorial Office

E-mail: energies@mdpi.com

www.mdpi.com/journal/energies

www.ingramcontent.com/pod-product-compliance
Lightning Source LLC
LaVergne TN
LVHW070610170726
843515LV00004B/37